SOTHIS PRESS

Sothis Press aims to completely reimagine the public domain book with an emphasis on universal access to captivatingly designed print editions. We believe that technology will enable every single book ever published to be perpetually accessible in beautiful print form. Sothis is beauty in ubiquity.

www.sothis.press
1 Waterhouse Square, London, EC1N 2ST

Small Sawmills

Their Equipment, Construction, and Operation

Daniel F. Seerely

SOTHIS PRESS

ISBN 978-1-0368-0818-1

This edition is published by
Sothis Press © 2023. The company
is registered in England and Wales.
The Company Number is 14900389.

UNITED STATES DEPARTMENT OF AGRICULTURE

BULLETIN No. 718

Contribution from the Forest Service
HENRY S. GRAVES, Forester

Washington, D. C. ▼ December 17, 1918

SMALL SAWMILLS

Their Equipment, Construction, and Operation

By

DANIEL F. SEEREY, Logging Engineer

CONTENTS

	Page
Object of the Bulletin	1
General Suggestions for Portable Sawmill Owners	2
The Mill and Milling	11
Logging	38
Appendix	57

WASHINGTON
GOVERNMENT PRINTING OFFICE
1918

UNITED STATES DEPARTMENT OF AGRICULTURE

BULLETIN No. 718

Contribution from the Forest Service
HENRY S. GRAVES, Forester.

Washington, D. C. ▼ December 17, 1918

SMALL SAWMILLS: THEIR EQUIPMENT, CONSTRUCTION, AND OPERATION.

By DANIEL F. SEEREY, *Logging Engineer.*

CONTENTS.

	Page.
Object of the bulletin	1
General suggestions for portable sawmill owners	2
Look before you leap	2
Capital required	3
Credits	3
Cost keeping	3
Organization	4
Mill site	5
Labor	5
Commissary	7
Marketing	7
Grading lumber	8
Auxiliary products	9
Some don'ts for sawmill operators	9
The mill and milling	11
Regular equipment	12
Auxiliary equipment	15
Engines and boilers	16
Water power	21
Belting	22
Circular saws	25
Log deck	31
Narrow gauge lumber lorry track	32
Setting up a portable mill	32
Operating the mill	34
Sawing	35
Piling lumber	37
Fire protection	37
Logging	38
Saw crew and equipment	38
Crosscut saws	39
Notching and felling	41
Preparing special products	42
The main logging road	44
Skidding	45
Skidways	46
Chutes	48
Loading logs	49
Scaling	51
Logging outfit	53
Appendix	57

OBJECT OF THE BULLETIN.

Running a portable sawmill is no longer an easy occupation. The more accessible timber in the West has mostly been cut out or burned, and to-day the principal stands are far back in the hills, making logging and milling expensive as well as strenuous work. Profitable operation calls for first-class logging equipment and modern mills, and for good business ability, skill, and hardihood on the part of the operator. Physical weaklings are more out of place in logging work than in any other kind of virile employment. Mere physical strength, however, is not in itself sufficient. A successful logger needs to be "strong" in the head as well as in the muscles.

This bulletin offers to portable sawmill operators suggestions regarding methods of organization, milling, and logging which have been proved by experience to give the best results. It is meant particularly for operators in National Forest timber, but should be useful to other owners of portable mills where conditions are like those in the National Forests.

GENERAL SUGGESTIONS FOR PORTABLE SAWMILL OWNERS.

LOOK BEFORE YOU LEAP.

Before purchasing a new mill or moving an old one to a new location, an operator should carefully inform himself regarding the following points:

- Amount of timber available for his operations.
- Kinds and grades of lumber the timber will produce.
- Density of the stand and conditions governing its cutting and removal.
- Method of marking timber to be cut.
- Method of scaling and lengths allowed for trimming.
- Penalty scale for broken trees and logs left on the ground.
- Length of time each year during which operations can be profitably carried on, and annual cut.
- Labor and supplies required and working capital necessary to finance the job.
- Margin of profit on which the business can be operated.
- Stumpage prices and how obtained.
- Payments and manner of making them.
- Logging costs and methods.
- Number of teams available for logging, for delivering lumber, and for hauling supplies.
- River and road improvements necessary.
- Brush disposal, and cutting and utilization of defective timber.
- Capacity, equipment, and power of mill.
- Milling or manufacturing costs.
- Distance to market and condition of roads.
- Capacity of market and prices for different grades.
- Competition.
- Possibility of establishing a small retail yard.
- Amount of lumber which market conditions require to be carried in stock.
- Market for by-products, such as mine timbers, railroad ties, telephone, telegraph, and power poles, field posts, cordwood.
- Responsibility for fire.
- Sanitary regulations.

If there is doubt about your ability to meet any of the conditions affecting the operation, go slow. It is a great deal better to find out beforehand that the chances are against success than to discover this fact only after you are in and can't get out.

CAPITAL REQUIRED.

Insufficient working capital is one of the prime causes for the failure of a small sawmill to make a reasonable profit. Even in a small operation considerable working capital is necessary for current expenses. Money must be available to pay for the equipment of the mill, the wages of the crew, the building of sheds, tracks, yard, bunk and cook houses, stables, and blacksmith shops, and for current expenses and repairs. Money is needed to carry a stock of lumber on the mill yard. These items of expense should be worked out in advance, and when money is not available at reasonable rates of interest, the prospective operator will be prudent if he resists the desire to become a sawmill owner. While it is true that a few have made a success from a very slender financial start, the chances of doing this are so remote that to embark in the sawmill business without the requisite capital is more of a gamble than a legitimate business proposition. A junked mill with an accompaniment of a few old axes and broken-down teams and harness, a rusty cross-cut saw or two, and, worst of all, several big bills and perhaps a mortgage, make a poor showing after several years of hard work.

CREDITS.

The unwarranted extension of credit is an almost universal practice among small operators, and usually results in disaster. The successful mill man sells for cash or negotiable paper. It is a common practice, however, to sell a load or two of lumber on time right along, even though the man who sells it owes his employees for labor or the merchant for supplies. If such a man is not mighty alert, he will soon find himself badly in arrears with his payments. Should that come about, the end of his career as a sawmill operator is in sight.

COST KEEPING.

The most important step in the operation of a sawmill is the opening of a simple set of books in which is recorded faithfully the cost of everything relating to the business. In the absence of such a record an operator is sailing on an unknown sea " without chart or compass." Yet, except in a very few cases, this necessary side of the business is entirely neglected. In a long experience with portable mill owners it was not until quite recently that the writer met one who gave it proper attention.

No elaborate system of cost keeping is needed by the small operator. All that is required is some simple form of accounting by which he can tell the value of his investments, the cost per thousand feet for logging and milling, the stumpage cost per thousand, the cost of repairs and new investments, the depreciation on logging and milling equipment, and the losses incidental to his business. Over against this should appear the value of the lumber sold by grade and the value of the yard stock by grade. Keeping a record of this kind is something that the average mill man is thoroughly competent to do himself, or he can get some member of his family to do it for him. A salaried bookkeeper is neither necessary nor desirable; he would cost too much. Once started on a simple system of cost keeping, the operator will, it is safe to say, be so much interested in the knowledge and insight which it gives him of his business that he will need no urging to keep it up.

While on the subject, it is pertinent to remind operators that hay, grain, vegetables, meat, etc., which are produced on their ranches and consumed on their logging operations should be charged against the sawmill account at the same price they would have cost if ordered from a storekeeper. The operator's own time, as well as that of his team while employed on the logging job, should also be charged to operating expenses. Very few sawmill men do this, the general idea seeming to be that if the business pays for the hired help and merchandise actually purchased from the store it is doing all that can be expected of it. Many operators seem to think that because they have made no money in the past, there is no possibility of making a better showing in the future. As long as operators have this feeling, it is morally certain that their condition will remain unchanged. Only when the men in the industry realize that they are engaged in a pursuit which calls for the best that is in them, and that increasing profits will reward their efforts, will the portable mill business take and hold its proper place among the staple industries of the country.

ORGANIZATION.

There are two ways in which an operator can organize his logging and milling work to obtain satisfactory results and at the same time know approximately how much each operation costs. One way is to keep the logging distinct from the milling, preferably carrying on the former during fall or winter, provided snow is not too deep. By this plan sufficient logs can be piled up in the mill yard or skidded up along the main road to keep the mill running during the season. The other and less desirable way is to take the entire crew into the timber and cut logs ahead for the season's run, afterwards working enough skidding and hauling teams to keep the mill supplied with logs.

The usual plan, however, is to log for a few days and mill for a couple more. This plan is neither economical nor efficient, for mill workers are very rarely good loggers and loggers are very rarely good millmen. Moreover, the axes, saws, and logging equipment get mislaid or are thrown aside after a few days' use. No one set of men is responsible for keeping them safe or in proper trim, and so valuable time is lost in assembling the equipment and getting it in working condition.

MILL SITE.

A number of points have to be considered in connection with the location of the mill. It should be near the water supply, but the buildings, and especially the toilets, should be so placed as to prevent any danger of the water becoming polluted. At the same time the mill should be at a point in the timber where there is from 500,000 to 1,000,000 feet of stumpage available for one setting, and where it will not be necessary to haul or skid the logs over long distances or uphill. Because a mill is small and portable is no reason why it should be moved very often, unless there is a good economic reason for doing so. Select a central site in the first place, where water and other conditions are favorable, and move only when the cost of moving, building new roads and camps, etc., can be saved by a shorter haul. The operator who moves his mill without figuring the attendant cost is likely to find that, though his mill may be small and portable, the expense of a new setting will not be small or very profitable in a financial way.

Yard and piling space must be provided for the lumber and slab piles, and a right of way for from 700 to 800 feet of narrow-gauge track on which to run the lumber from the mill to the yard. There must be a landing deck for logs, with log decks and skidways, and provision must be made for the economic handling of sawdust and bark. Convenient locations must be found for the bunkhouse and the camp dining room and storeroom. Toilets and covered refuse pits are other essentials.

A rough ground plan of the proposed plant will materially assist the operator in selecting sites for the different buildings. In some instances tents can be substituted for wooden structures. Dry wood for domestic use is sometimes an important consideration. The entire area round the mill and buildings should be cleared of brush and débris and kept cleared. Two or three milch cows and a few hogs and poultry can be maintained around a small mill at very little expense

LABOR.

It is essential for the success of a small mill operation that the logging crew should be made up of experienced men. Green hands may succeed fairly well around the mill handling lumber or sawdust, but the woodsmen must be trained or else they will not do enough effective work to pay for their board. Green hands attempting to fell, skid, load, and haul logs are only about 25 per cent efficient, without reckoning the loss from broken timber, split trees, etc., which if taken into account would reduce their efficiency close to zero. The custom of trading lumber for inexperienced labor, so prevalent in many places, should never be practiced by the small operator.

A real knowledge of how to handle horses is very essential in logging. Thousands of dollars' worth of good horseflesh is ruined by ill-tempered, incompetent teamsters. No part of the operator's investment needs closer supervision than the hauling. A poor teamster is poison (I know of no more expressive term) to a good team. An operator will save money by keeping his horses idle in the barn, no matter how badly their services are needed, rather than allow a brainless teamster to pound them through the timber and over rocks, stumps, and mud holes. Discharge such a person at once. Teams when properly handled will be 100 per cent efficient all the time and thrive.

High wages do not always secure the service of competent labor, particularly in woods work. One gang of sawyers may cut the same amount of board feet in logs as another gang, and yet may cut their logs with such a disregard of correct lengths and of crooks and with such an indifference as to how the trees are felled for skidding purposes that the value of their labor may be only 50 per cent as much as that of the other crew, who do their work as it ought to be done.

The millman who neglects to supervise his operations rigidly is surely preparing the way to financial disaster. An operator usually works hard at some particular job, such as sawing, and leaves the rest of the work to run itself. His proper place is "bossing the job," and if he does that thoroughly he will have his hands full. In order to instruct men in woods work, the operator must understand it himself. If he lacks this knowledge it would be wise for him to keep out of the portable-mill business or else hire a competent man to run it for him.

Clean, wholesome living and sleeping quarters for the men, as well as properly cooked food, deserve close attention—a good deal more attention, in fact, than is usually bestowed upon them. "Sour dough" grub and rough living may sound very romantic in a cheap novel, but when actually practiced they fail to bring results. Pay the men good wages and feed them well, and see that they earn it. Always have money on hand to pay off men if they quit or you discharge them.

Make it very plain that you want value for every cent you pay in wages or in board. Tolerate no "deadheads" around your camp. Be boss yourself or delegate the job to some one who is qualified to fill the bill. Do not make your camp a dumping ground for all your male relatives. You are supposed to be running a sawmill—not a rest cure.

In short, inject a lot of energy and vigor into the business, or leave it to some one who can.

COMMISSARY.

Every operator should keep a small supply of dry goods in stock for the use of his men, and also such articles as tobacco, matches, pipes, medicine, and writing pads and pencils. Old magazines and books are easily obtained and are eagerly read. The supplies enumerated should be bought in bulk or by wholesale, in order to cover the cost of freight and handling. Kerosene and lubricating oils should be purchased by the barrel, and enough camp supplies, such as flour, feed, pork, potatoes, hay, groceries, and canned stuff, should be kept in stock to avoid the necessity of sending a man and team to town every week. The item of supplies is an important one to the success of the job and should be handled economically.

MARKETING.

Every town and village on a railroad has at least one retail lumber yard, and very often more than one. It is one of the first institutions to be established after the advent of the railway. In towns and villages remote from the railway, however, lumber yards are seldom in evidence, although there may be and usually are half a dozen portable mills hidden away in the hills in the vicinity. Yet a lumber yard is as much a necessity and could command as much business in many of these remote communities as the one or two yards in the railway town.

Right here is an opportunity for wide-awake portable mill owners to establish themselves in the lumber business. The surprising thing is that so few have taken advantage of it. If a rancher needs a load of lumber and sees a yard right in town he will buy material and take it home with him; whereas if he knows that he has to drive to the mill 10 or 12 miles in the hills over a wretched road, and is not sure of getting what he wants when he arrives there, he will naturally put off a disagreeable job as long as he can. Eventually, if he has business in the railroad town, he will purchase from the retail yard there. An occasional short item in the local paper telling the public something about the activities of the local sawmill and the amount of lumber turned out daily, with the stock on hand, makes interesting reading for the local public, particularly for people who need lumber.

There are scores of towns and villages in the inter-mountain country where such opportunities exist to-day. Yard space would cost very little and teamwork is cheap, and the millman who takes a little pains in milling and grading his lumber and putting it up in neat piles will soon have his mill running steadily to keep up with the demand. There is little to be gained in sawing out a lot of lumber and keeping it hidden away in the hills. For every man, woman, and child in the United States 375 feet of lumber is used annually; in new States three or four times as much. Montana, for example, uses 1,234 feet per capita, and it is a moderate estimate that a community of 500 people will use up a quarter of a million feet annually. A small mill operator who opens a yard and keeps 100,000 feet of lumber in stock, along with a moderate amount of shingles, lath, and building material, can establish a remunerative business very easily. Be it understood that lumber in this sense means lumber that is properly sawed, surfaced (when necessary), edged, trimmed, graded, air dried, and properly piled. You can not run a successful lumber yard with rubbish anywhere.

The heaviest demand in newly settled communities is for low-grade lumber, viz, No. 3 common and dimension, the very class of material which the portable mills can supply most readily. The upper grades will always find a remunerative market. A small mill, if properly handled, can successfully hold a competitive market for low-grade lumber and dimension against similar products produced in a large mill and shipped into the local market.

GRADING LUMBER.

The necessity for grading the product of small mills can not be emphasized too strongly or too often. Every millman is able to distinguish between good and bad, rotten and sound lumber, and what sort of a log is best adapted for inch lumber and what for plank, dimension, finish, and so on. But when it comes to being able to tell at a glance what defects in a board causes it to grade No. 1 common instead of inch finish very few can make an intelligent distinction, yet those men have been handling such lumber perhaps a lifetime. They have simply neglected to use their powers of observation.

The Western Pine Manufacturers' Association of Spokane, Wash., issues free a little booklet containing the rules for the grading of pine, fir, and larch, which is distributed by their secretary to anyone interested. To a man handling lumber every day this little book will be invaluable. To have lumber to sell and no grading rule to sell it by is a condition that spells certain loss for the sawmill owner. If he doesn't study grading himself, his son will, with the result that the boy will learn more about the lumber business in a month than his father has been able to pick up in years. There is no "royal road" to a knowledge of lumber grading—you must learn it yourself.

It sometimes happens that when an enterprising mill operator, who has adopted modern methods in milling and grading his output, is negotiating the sale of a bill of lumber with a prospective purchaser, another operator with an old mill up in the hills will jump in and take the order away from him at a lower price. That particular bill of lumber, it is safe to predict, will be sawed in any old way, with the result that the purchaser is dissatisfied with the material he gets, and concludes that native timber is no good, and never buys any more of it. By this sort of thing the unprogressive mill operator spoils a lot of good lumber at no profit to himself. It doesn't pay to drive away future trade for the sake of a single sale. In these days men no longer have to use the roughest kind of lumber for lack of something better. They may use it once, but it is a moral certainty that they won't come back for any more. The way for an old-time operator to get and hold trade is to turn to modern methods, not to try to put another and more progressive man out of business. A new order prevails in the lumber business, and no one knows better than the small millman that there is no profit in logging along the old lines. A determined effort must be made to get out of the old rut and place the portable mill industry on a paying basis.

AUXILIARY PRODUCTS.

The demand for lumber and dimension stuff—the regular products of the mill—is always supplemented by a demand for the more finished products, such as planed lumber, siding, flooring, lath, shingles, etc., which require special machinery for their manufacture. The millman will find that he can get more for his best grades of lumber if he can furnish them as required by the local trade. Considerable waste may be saved in the utilization of short lengths and slabs, which otherwise would be lost. There is money in reducing waste when it can be partially transformed into a salable product.

SOME DON'TS FOR SAWMILL OPERATORS.

Before outlining in detail the necessary equipment for a portable mill and how it should be operated a few suggestions as to what to avoid or, in other words, what not to do may not be amiss To mill operators:

1. If you have a market and a cutting capacity of 500,000 feet per year, do not enter into a contract with the Government or with anyone else to cut 5,000,000 feet in three years.

2. If you are financially unable to make more than $300 in advance payments, do not sign a contract to pay $500.

3. Do not sign a contract with the Government to cut timber on a National Forest without carefully going over the area and finding out for yourself what species of timber you are expected to cut and where it is located.

4. Do not try to fell timber with a dull, rusty cross-cut saw and wooden wedges. It won't pay.

5. Do not try to keep your axes sharp and serviceable with a rusty file. Try a grindstone.

6. Do not try to skid saw logs without skidding tongs, swamp hook, cant hook, and chain. Handspikes and hand-beam levers are out of fashion.

7. Do not try to haul saw logs on a lumber wagon. Try a heavy truck or a logging sled if you have sufficient snow.

8. Do not try to haul saw logs uphill or over rocks and brush. Build a road.

9. Work steadily and systematically. If half your crew leaves on Saturday and does not return to work until the following Tuesday, get steadier men. Blow your whistle at 7 o'clock every morning and get to work at 7, not 9.

10. Do not unload your saw logs all over the mill yard and then roll them over rocks, boulders, slabs, brush, and mud holes with a handspike. Build a downhill skidway in front of the carriage, and the logs will roll down by gravity.

11. If your mill can manufacture 10,000 feet per day, see that it does that every day you run it. Capacity cutting for one day during the season does not count.

12. If your boiler is old and leaky, get it repaired and inspected. A rusty, leaky boiler is likely to stop all business quite suddenly. Watch the water gauges.

13. Use dry slabs and clean water in making steam. Green slabs, dirty water, and a leaky boiler make a bad combination. Use the mud cock frequently.

14. Do not try to manufacture merchantable lumber with a dirty, rusty engine set on a rotten foundation with a shaky mandrel, rotten belting, a saw out of "true" and running at half speed under insufficient steam from a leaky boiler. It can't be done.

15. Do not expect your mill to be a profitable business if you leave your mill machinery exposed to the weather the year round. You will soon have a junk pile, not a sawmill.

16. Do not expect to get the value of your lumber if you throw it in a heap in the mill yard instead of piling it properly and grading it.

17. Do not sell your lumber on credit or on time. You have not sufficient capital to do a credit business.

18. Do not keep your accounts on a shingle or marked up on a board. A small Dr. and Cr. account book is much more satisfactory.

19. Do not try to run a sawmill, however small, by rule-of-thumb, hit-or-miss methods that your grandfather practiced. They will not work in this day and age.

20. Do not forget that this is the twentieth century and that the management and successful handling of a portable mill is a twentieth-century problem.

THE MILL AND MILLING.

If the operator is already provided with a mill, the size of his operations will depend on the power and capacity of that mill. If he plans to purchase a new mill, he should figure on one with the power, capacity, and equipment necessary to meet the requirements of his capital, his market, and his logging equipment. Lack of sufficient power, even for driving the saw, is a common handicap for small mills. An effective 25-horsepower engine will, if properly handled, keep a circular saw working to capacity and at the same time will run a small planer and edger.

The type of mill in most general use in Forest Service District 4 is one cutting from 2,000 to 20,000 feet per day. The following discussion, therefore, will be confined to two classes of this type, one cutting from 2,000 to 10,000 feet per day and a heavier one cutting from 10,000 to 20,000 feet per day. The smaller mill will be called Class A, and the larger, Class B.

Sawmills are generally classified as right and left hand mills, according to whether the log passes to the right or to the left of the saw, viewed from the front. In ordering equipment for an old mill it is always necessary to specify whether the equipment is needed for a right or left hand mill.

Portable mills are usually equipped with rack and pinion or cable drive, and friction or belt feed or a combined belt and friction feed. Shotgun feed is not used in small mills. The variable friction feed is so called because the sawyer can vary the feed to correspond with the power or the size and species of timber to be cut, easing down on knots and frozen timber and increasing the feed at will. With the same power a variable friction-feed mill is said to cut from 25 to 40 per cent more than a belt-feed mill. There are no belts or springs to break or give trouble. The belt feed is a combination of belt and flat face frictions so arranged that the feed and gig back frictions are continuously driven in opposite direction by an endless belt direct from the mandrel. A single lever operates the feed and gig back by shifting the bull wheel from one friction to the other. The shaft of the bull wheel has a pinion on the opposite end which operates the wire cable drum, doing away with all intermediate gears and securing a strong direct drive for the carriage. The friction shafts have babbitted boxes with screw adjustment to take up the wear in the frictions, and a substantial idler is provided to keep the endless feed belt always tight. Wire cable or rack and pinion drum can be used on both belt-feed and friction-feed mills.

"Never buy anything because it is cheap" is a form of advice that applies very particularly to a sawmill. Going a little further, it might be laid down as an excellent business maxim: Never buy an old second-hand mill, no matter how cheap it can be purchased. Regardless of type or make, however, a mill will turn out to be a liability rather than an asset unless it is set up properly and kept in good running order, and the work conducted on a business basis.

The word "portable" as defined in the dictionary means "capable of being borne or carried, easily transported, conveyed without difficulty." The average portable mill owner, however, seems to take this definition too literally, for when he purchases a new mill, or moves to a new site, he pays as little attention to setting it up or lining the machinery properly as he would in the case of a thrashing engine or a steam plow. The usual procedure is to clear off a level space of ground, set down the engine and boiler and block them up a little, adjust the belting, get up steam, and start the saw. The moving and setting up is not taken seriously. Regardless of the length of time a portable mill is to remain on the same site, the engine bed should be solid, the belting properly adjusted, the engine, sawmill, and carriage properly lined up and housed and all working parts oiled and the shafting in line and all parts in good working order. Otherwise it is impossible to manufacture good lumber.

In general, the life of a portable mill may be placed at from 12 to 15 years. One hundred and fifty days is the maximum time it is in active operation each year, so that, if not properly cared for, a portable mill may be said to "rust out" faster than it wears out.

REGULAR EQUIPMENT.

CLASS A.

Following are the various parts and equipment of a class A mill complete and their cost in 1916. Since the entrance of the United States into the war, prices have fluctuated to such an extent as to make it impractical to give those likely to be in effect when this bulletin is issued. It has been thought best, therefore, to give 1916 prices throughout as a basis for estimates.

Sawmill (husk and carriage)	$254.00
Sawdust conveyors, 25-foot	35.00
48-inch chisel-tooth circular saw, 8 gauge	62.00
60-inch solid-tooth circular saw, 7 gauge	82.00
Engine and boiler, 15 horsepower	776.00
Pony edger	145.00
Trimmer	280.00
Shingle machine	160.00
Lath mill and bolter	200.00
Shingle buncher and packer	10.00
Axes, cant hooks, shovels, wheelbarrows, etc	20.00
Chisel tool files, dozen	5.00
Total	2,029.00
Discount 20 per cent	405.80
	1,623.20

The equipment of the sawmill usually consists of—

Husk frame, 7 by 3 feet, built of 3½ by 7½ inch timbers.
Variable-friction feed.
Steel mandrel, 2¾ inches by 4 feet 4 inches.
Board roll and spreader wheel and saw guide.
Carriage 16 feet long and 26 inches wide, of 3½ by 5½ inch timber.
Feed rack, 22 feet long.
Four trucks with 6-inch wheels and 1⅛-inch steel axles.
40 feet of V and flat rolled steel track.
Two head blocks, opening 34 inches, with duplex dogs.
Ideal set works with quick receder.
14-inch polished-steel set shaft.

There is also furnished a mandrel pulley up to 24 inches diameter, belt tightener, foundation bolts, cant hook, oil can, and wrenches for rack and pinion-carriage drive. Also trackway timbers bolted together with steel track attached.

The price of a sawmill so equipped, as shown in the preceding cost statement, is $254 f. o. b. Chicago. The manufacturers allow 15 per cent discount on time, and 20 per cent for cash. The total shipping weight is 2,550 pounds. Way timbers are 450 pounds extra.

Extra equipment:

Manila-rope drive with sheaves, add	$10.00
Wire-cable drive, add	25.00
For each additional foot of carriage, add	3.50
For each additional foot of set shaft, add	.60
Trackway timbers, if not wanted, deduct	16.00
Belt tightener, if not wanted, deduct	13.00

A top-saw attachment is too heavy for this type of mill. All mandrels are made for saws with 2-inch center hole and two ⅝-inch pinholes on 3-inch circle, and are adapted to any power from 6 to 15 horsepower. This outfit can cut 2,000 feet per day with a 6-horsepower engine, or from 7,000 to 8,000 feet per day with a 15-horsepower engine. It will carry a saw up to 52 inches and logs up to 36 inches diameter.

CLASS B.

The class B mill is heavier and of larger capacity than the class A mill, and has an independent rack beam near the center of the carriage. It is adapted to engines of from 15 to 40 horsepower, and will cut from 10,000 to 20,000 feet per day. Saws up to 60 inches may be used. The headblocks will open to receive a log as large as 51 inches in diameter. The parts and equipment are

Sawmill	$151
Sawdust conveyors	35
48-inch inserted-tooth circular saw, 8 gauge	62
60-inch solid-tooth circular saw, 7 gauge	82

Engine and boiler, 25 horsepower	$1,058
Pony edger	145
Trimmer	280
Planer	169
Slash saw	60
Shingle machine	160
Lath mill and bolter	200
Shingle buncher and packer	10
Tools, axes, cant hooks, shovels, etc	20
Inserted-tooth files, per dozen	6
Total	2,741
Discount 20 per cent	548
	2,193

The equipment of the sawmill consists of—

Husk frame 8 feet 6 inches by 4 feet built of $4\frac{1}{2}$ by $11\frac{1}{2}$ inch timbers.
Variable-friction feed.
Steel mandrel $2\frac{15}{16}$ inches by 5 feet 6 inches long.
Mandrel pulley 24 by 12 inches.
Board roll, spreader wheel, and saw guide.
Carriage 24 feet long, 40 inches wide, timbers $5\frac{1}{2}$ by $5\frac{1}{2}$ inches.
Feed rack 32 feet long.
Six trucks with 10-inch wheels and $1\frac{11}{16}$-inch steel axles.
Fifty-six feet of V and flat rolled steel track.
Two head blocks opening 48 inches with Grant duplex dogs.
Twenty feet polished-steel set shaft.
Trackway timbers framed and bolted together in sections with steel track attached, belt tightener, foundation bolts, cant hook, oil can, and wrenches.
Rack-and-pinion carriage drive.
Saw not included. Weight 5,250 pounds.

The extras are—

Manila-rope drive with sheaves, add	$10
Wire-cable drive with carriage, add	36
Carriage, each additional foot, add	6
Set shaft, each additional foot, add	1
Top saw attachment, add	80
Trackway timbers, if not wanted, deduct	20
Belt tightener, if not wanted, deduct	14

Sawdust conveyor fixtures (weight, 142 pounds).
Top mandrel is $1\frac{15}{16}$ inches diameter and made for standard saws with 2-inch hole and $\frac{5}{8}$-inch pin holes on 3-inch circle. Weight, 1,075 pounds.

The average cost of setting up the mill is—

Mill (with machinery on the ground)	$300
Housing	250
Boarding and sleeping camps	200
Total	750

AUXILIARY EQUIPMENT.

EDGERS.

A 33-inch pony gang edger with a capacity of from 5,000 to 11,000 feet per day has two saws with single-pressure roll, and cost in 1916—

With solid-tooth saws	$145
With inserted-tooth saws	175

A discount of 40 per cent is allowed from these prices. Specifications for edgers of this size are—

Floor space, 18 feet 6 inches by 48 inches.
Width inside, 33 inches.
Guide adjustment, 4 inches.
Mandrel pulley, diameter, 1⅞ inches.
Mandrel pulley, dimensions, 8 by 8 inches.
Saws, diameter, 14 inches.
Speed, revolutions per minute, up to 2,000.
Feed belt, 8 feet long, 4 inches wide.
Maximum opening between saws: Two-saw edger, 24 inches; three-saw edger, 20 inches.
Weight: Single roll, 1,300 pounds; double roll, 1,400 pounds.

PLANER.

Planers are built in two sizes, to surface up to 16 inches and 20 inches wide and from ¼ inch to 6 inches thick, with or without countershaft, as ordered. No belting is furnished. Feed belts require 13½ feet of 1½-inch belting.

One pair planer knives, weight 745 pounds:	
16-inch machine with countershaft	$152
20-inch machine with countershaft	169
Pony planer, weight 1,700 pounds:	
20-inch	275
24-inch	325
Planer and matcher	360

Prices are subject to discount of 20 per cent, Chicago.

SAWDUST CONVEYORS.

Sawdust conveyors are labor-saving devices that constitute a good investment. The standard 25-foot conveyors consist of the necessary sprocket wheels, bevel gears, sprocket and gear shafts, boxes, collars, 50 feet of carrier chain with conveyor attachments, and 10 feet of drive chain.

Price of 25-foot conveyors as described	$35.00
Price per foot extension	$0.15
Longest run recommended ... feet	60
Approximate weight ... pounds	150

TWO-SAW TRIMMER.

No. 1 machine, for boards from 6 to 16 feet, weight 2,650 pounds	$280
No. 2 machine, for boards from 6 to 18 feet, weight 2,750 pounds	300
No. 3 machine, for boards from 8 to 20 feet, weight 2,850 pounds	320
No. 4 machine, for boards from 8 to 24 feet, weight 3,110 pounds	340

CUT-OFF SAW.

Timbers in main frame 3½ by 3½ inches.
Timbers in table frame 2 by 4 inches.
Mandrel (steel) 1$\frac{1}{16}$ inches for saw with 1¼ inch hole.
Mandrel pulley 4 by 4 inches.
Size of saw, 20 inches. (Any size can be used from 18 to 24 inches.)
Floor space 4 by 10 feet.
Shipping weight, 430 pounds. Price f. o. b. Chicago, $70. Iron parts only, no saw. Discount 22 per cent.

LATH MILL AND BOLTER.

Frame, 8 feet long, 37 inches wide, and 27 inches high.
Lath, saw, 10 inches; bolter saws, 22 inches.
Driving pulley on each mandrel, 8 by 8 inches.
Speed of lath saws, 2,700; bolter, 1,400.
Weight of combined machine, 1,050 pounds.
Complete with 3 lath saws and bolting saw, $200. Discount 22 per cent, f. o. b. Chicago.

GANG LATH MACHINE.

3 saws	$130
4 saws	134
5 saws	138
6 saws	142

20 per cent discount f. o. b. Chicago.
Frame, 37 inches long, 37 inches wide, and 27 inches high.
Lath saws, 10 inches.
Driving pulley on mandrel, 8 by 8 inches.
Speed of saws, 2,700 r. p. m.
Weight, 675 pounds, net.

SHINGLE MACHINE.

Cuts 16 to 18 inch shingles with 36-inch saw, $160; 20 per cent discount, f. o. b. Chicago.

SHINGLE BUNCHER OR PACKER.

Size of bunch, 20 inches wide, 30 inches long.
Price, $10. Weight, 80 pounds.
Discount, 20 per cent f. o. b. Chicago.

ENGINES AND BOILERS.

TYPES AND CAPACITIES.

The usual specifications for a 15-horsepower engine and boiler suitable for a class A mill are—

Size of cylinder, 8 by 10 inches.
Revolutions per minute, 100.

Diameter of pulley, 44 inches.
Face of pulley, 10 by 10½ inches.
Diameter of boiler, 32 inches.
Length of furnace, 44 inches.
Height of furnace, 33 inches.
Width of furnace, 26 inches.
Number of 3-inch tubes, 26.
Length of tubes, 78 inches.
Shipping weight, 7,300 pounds.

The price is $776. The following are furnished as regular equipment: Oil cups, sight-feed lubricator, steam gauge, water gauge, whistle, gauge cocks, throttle, blow-off check, stop and safety valves, smoke stack, grates, governor, belt pulleys, pet cocks and wrenches, stack and guy wires, steam and exhaust connections and injectors.

The center crank horizontal engine mounted on a water front open bottom locomotive boiler is suitable for class B mills. The engine is firmly bolted to brackets securely fastened to the shell of the boiler. It can also be detached and used separately.

The fittings and fixtures comprise grates, water column, water gauge, gauge cocks, steam gauge, safety valve, check and stop valves, whistle blow-off valve, stock and guy rods, injector attached, fly-wheel pulley, oil cups, sight-feed lubricator, throttle valve and nipples, automatic governor and belt, pipe connection between boiler and engine, and exhaust pipe; pump and heater instead of injector and spark arrester extra.

The usual specifications for a 25 or 30 horsepower engine attached to locomotive boiler, suitable for class B mills, are—

	25 horse-power.	30 horse-power.
Size of cylinder ... inches..	10 by 12	10 by 12
Revolutions per minute	160	190
Diameter of pulley ... inches..	24 and 52	24 and 52
Face of pulleys ... do....	12 and 12½	12 and 12½
Diameter of boiler ... do....	36	36
Length of furnace ... do....	52	52
Height of furnace ... do....	[illegible]	40
Width of furnace ... do....	30	30
Number of 3-inch tubes	34	34
Length of tubes ... inches..	96	120
Diameter of smokestack ... do....	16	16
Length of smokestack ... feet..	25	25
Size of steam pipe ... inches..	2½	2½
Size of exhaust pipe ... do....	3	3
Diameter of shaft ... do....	3¾	3¾

The water-front, open-bottom boiler has proved to be one of the best types. With the open bottom there is no dead water space below the grate surface, such as exists in the water-bottom boilers, to fill with mud and sediment. The open bottom permits the free discharge and easy removal of ashes, thus providing free draft, an important consideration in the case of the poorer grades of fuel such as are used in mountain mills. The fittings and fixtures comprise grates, water column attached with gauge cocks, water gauge, steam gauge, safety valve, check valve, stop cock, whistle, blow-off valve, stack and guy rods (four times the length of the stack); injector and spark arrester extra.

The usual specifications for a 25 or 30 horsepower boiler are—

	25 horsepower.	30 horsepower.
Diameter of boiler ... inches..	36	36
Length of fire box ... do....	52	52
Height of fire box ... do....	38	40
Width of fire box ... do....	30	30
Number of 3-inch tubes ...	34	34
Length of tubes ... inches..	96	120
Thickness of shell ... do....	1/4	1/4
Thickness of furnace plates ... do....	9/32	9/32
Thickness of tube sheets and heads ... do....	3/8	3/8
Size of dome ... do....	20 by 24	20 by 24
Diameter of stack ... do....	16	16
Length of stack ... feet..	25	25
Number of steel in stack ...	16	16
Weight of bare boiler, on skids ... pounds..	5,300	5,800
Weight, with fixtures attached ... do....	6,300	6,900

The following is a list of the component parts of a 25 and 35 horsepower center-crank horizontal engine and the approximate cost of replacing each part:

	25 horsepower.	35 horsepower.
Size of cylinder ... inches..	10 by 12	10 by 13
Bed frame ...	$114.00	$114.00
Connecting rod, complete ...	58.20	64.40
Crank shaft, complete ...	96.00	116.00
Crank pin, brass ...	13.00	14.00
Crosshead-pin, brass ...	7.50	7.50
Crosshead, complete ...	26.52	26.52
Cylinder head, plain ...	13.20	15.60
Cylinder head, with gland ...	19.80	24.00
Cylinder, complete ...	69.60	78.00
Eccentric rod ...	5.40	6.12
Eccentric rod head ...	7.68	7.68
Eccentric and strap ...	12.00	13.68
Flywheel ...	46.80	46.80
Governor pulley ...	6.12	6.60
Piston ring, each ...	2.40	2.70
Piston ring with rod fitted ...	15.72	18.24
Piston-rod gland ...	3.48	3.48
Pulley ...	13.90	13.90
Steam-chest cover ...	8.34	8.34
Valve ...	3.72	3.84
Valve stem ...	6.54	6.72
Valve-stem gland ...	4.86	4.86
Valve-stem bar ...	7.26	7.23

CARE OF BOILER.

The first duty of an engineer when he enters his boiler room in the morning is to ascertain how many gauges of water there are in the boiler. Never unbank or replenish the fires until this is done. If the water is low, cover the fires with ashes or fresh coal and close the drafts and ash-pit doors. Do not turn on the feed or tamper with or open the safety valve. Let the steam outlets remain as they are. In really dangerous cases draw the fires.

In cases of foaming, close the throttle and keep it closed long enough to show the true level of the water. If the level is sufficiently high, feeding and blowing will suffice to correct the evil. In cases of violent foaming caused by dirty water or a change from salt to fresh water, or vice versa, check the drafts and bank the fires in addition. If leaks start in the boiler, repair them at once.

To blow off, clean the furnace and bridge of all ashes and débris. Allow the fire box or brickwork, as the case may be, to cool down for an hour or two before opening the blow-off. The pressure should not exceed 20 pounds when a boiler is blown out. Blow out at least once or twice in two weeks. In case the feed becomes muddy, blow out some every day. When surface blow cocks are used they should often be opened for a few minutes at a time. After blowing down, allow the boilers to become cool before filling up again. Cold water should never be pumped into a hot boiler; it will cause sudden contraction in the plate. In tubular boilers the handholes should often be opened and all dirt and sediment removed. When boilers are fed in front and blown off through the same pipe the mud and sediment in the rear end should be cleaned out. Raise the safety valves cautiously and frequently; otherwise they are likely to become fast in their seats and useless.

Should the gauge at any time indicate the limit of pressure allowed, see that the safety valves are blowing off. Keep the gauge cocks clear and in constant use. Glass gauges are not reliable. In preparing to get up steam after the boiler has been open or out of service, great care should be taken in making the manhole and handhole joints. The safety valve should then be blocked open, and the necessary supply of water run or pumped into the boilers. In tubular and locomotive boilers this should be until the water shows at second gauge; in vertical tubulars a higher level is advisable as a protection to the top end of the tubes. After this is done fuel may be placed on the grate, dampers opened, and fires started. If the chimney or stack is cold and does not draw properly, burn some oily waste or light chips at the base. When steam issues from the safety valve, lower the valve carefully to its seat, and note pressure and behavior of steam gauge. Under all circumstances keep the gauges, cocks, etc., clean and in good order and things generally in and about the engine and boiler in a neat condition. When a blister appears on the boiler there must be no delay in having it carefully examined and trimmed or patched. Particular care should be taken to keep the sheets and parts of the boiler exposed to the fire perfectly clean, and all tubes, flues, and connections well swept. This is particularly necessary where wood is used for fuel.

The extent of the heating surface of a boiler depends on the length and diameter of the shell and the number and size of the flues. It is customary in calculating the heating surface of the shell to consider that two-thirds of it and the entire surface of the flues is exposed to the action of the heat.

STEAM BOILER WATER.

A steam boiler needs good water as much as it needs good fuel. All water used in boilers contains more or less impurities and acid. Impurities which cause trouble include soluble salts of calcium and magnesium, bicarbonates of alkaline earths, and sulphate of lime. Water containing more than 0.005 per cent of free sulphuric or nitric acid is likely to cause serious corrosion, and more than one-tenth of 1 per cent of acid will cause scale. Hard water invariably forms scale, and comparatively soft water may also do so if the boiler is used too long without being emptied. Foaming is caused chiefly by an excess of alkaline salts, which causes the water to form suds as if soap had been placed in it.

Scale not only decreases the efficiency of the boiler but also causes deterioration; for, when it is sufficiently thick the conducting power of the boiler is reduced, and the tubes and plates become overheated and crack or burst. Again, the scale may keep the water from sections of the heated plates for some time, and then give way, causing large volumes of steam to be suddenly generated, possibly resulting in an explosion. It has been demonstrated that $\frac{1}{8}$ of an inch of scale in boilers causes a loss in heat transmission of from 10 to 12 per cent, and this loss of heat increases with the thickness of the scale. A porous scale retards the heat transmission more than a solid scale. It sometimes happens that different kinds of scale may be found in the same boiler, owing to the different temperatures of the sheet in different parts and to the circulation of the water. The scale on the tubes is also different from that on the sheets, owing to the same causes.

Methods for the purification of feed water consist of the use of feed-water heaters, scum catchers, and blow-off valves, or chemicals placed in the boiler or in the water before it reaches the boiler. Purifying chemicals placed in the boiler are soda ash, caustic soda, phosphate of soda, tannin compounds, fluoride of soda, and aluminate of soda. As a rule, the expense of purifying feed water with chemicals makes it prohibitive for small mills. The use of feed-water heaters, scum catchers, and blow-off valves, however, is strongly recommended.

To prevent the adherence of scale to the boiler shell, many substances have been used, such as potatoes, kerosene, and other remedies, organic and mineral. Of the boiler compounds found on the market none have given more general satisfaction than those which have soda and some form of tannic acid as their base. Tannic acid has a slight action on the iron of the boiler and is reasonably efficient in preventing scale from sticking.

To prevent scale, or at least to diminish the injury it does to boilers, the feed water should be heated by live steam under boiler pressure in a separate vessel before going to the boiler. By allowing the water to settle for some time, sediment, mud, and dirt will be got rid of. Oil from the engine cylinder is particularly injurious to boilers, and, when noticed in the condensed steam, should be carefully removed.

STEAM PUMPS.

If water can not be obtained under sufficient pressure to cause it to flow into the intake by gravity, some sort of pump is necessary. Usually the pump runs by steam. If the pump runs badly, make sure the water valves and water pipes are all right before examining the steam end. When the pump is not in use in cold weather open all the cocks and drain plugs to prevent freezing. Always see that the pump has a full and steady supply of water to work on. Do not take the pump apart to see what is inside as long as it does its work well. Set a pump on a level solid foundation so as to avoid undue strain on the pipes and resulting leaky joints. Long pipes should be larger than short ones, to allow for increased friction. All pipes should be as short and straight as possible. A foot valve and strainer should be attached to the suction pipe.

Use few elbows, tees, and valves, substituting full round bends for elbows, and wyes for tees; sharp bends greatly increase the friction. Care must be taken to guard against leaks in the suction pipe, as a very small leak will supply the pump with air to its full capacity, and little or no water will be obtained. A suction air chamber made of a short nipple, a T, a piece of pipe smaller in diameter than the suction pipe and from 2 to 3 feet long, and a cap, screwed upright into the suction pipe close to the pump are always useful, and when the suction pipe is long, in high lifts, or when the pump is running at high speed, it is a positive necessity. Its use insures a steady and uniform flow of water through this suction pipe and prevents "pounding" or "water hammering."

Use good cylinder oil, and oil the steam end just before stopping the pump. Keep the stuffing boxes full of good packing, well oiled, and just tight enough to prevent leakage without excessive friction.

WATER POWER.

As a general thing, water power is not practicable for a portable mill. It may be found in one location, but not in the next. It is cheaper than steam, but the source of power is neither uniform nor continuous; it is not as capable of control as is steam, and in most cases it gives low speed and low capacity. Decriptions of various types of water wheels will be found in the Appendix.

BELTING.

Belts transmit power from the engine to the saw. The heavier the belt the more power it transmits. Belt tighteners are required when a belt itself is not heavy or long enough to cause sufficient sag. The sag should always be on top, in order to increase the arc of contact with the pulley. Belts are made either of leather, rubber, or fabric. Leather belts are sold either single or double. The transmitting power of a single belt is only 70 per cent of that of a double belt. Rubber belts stand moisture better than leather and are cut from $\frac{1}{8}$ to $\frac{1}{4}$ inch shorter per foot than the circuit on which they run. They are run with the seam side out, while leather belts are run with the grain side in. Rubber belting is sold as 2, 4, 6, and 8 ply, the 4 ply being equal to single leather belting, the 6 ply to light double leather belting, and the 8 ply to heavy double leather belting. Small mills are usually equipped with either fabric or rubber belting.

When ordering belting of any kind it is necessary to specify (1) diameter of driving pulley, (2) its revolutions per minute, (3) diameter of driven pulley, (4) distance of pulley centers, (5) horsepower to be transmitted, and (6) width of pulley face.

Leather belting is spiked or joined either by studs or by belt cement. Next best to the cemented joint in a leather belt is that made with a rawhide or other lace. But this joint in a leather or any kind of belt must be made properly. Large lacing holes and a big bunch of lace may be almost as harmful and cause as much "jump" in the belt as a double row of studs. Where leather belting is exposed to moisture and waterproof cement is not accessible, lacing may be resorted to; but, as a general rule, all leather belts should be cemented.

In lacing a fabric belt never use a hollow punch, because it cuts the stitches and weakens the belt. Use a pointed awl. Cut the ends of the belt perfectly true with a try-square and punch or bore two rows of holes in each end. The holes in the second row should be punched directly back of the holes in the front row. The holes in both ends of the belt should also be directly opposite. No hole should be less than one-half inch from either edge or end of the belt. The holes should be spaced three-fourths inch from center to center. When an odd number of holes are to be laced, begin with the center hole; and when an even number, start with either of the two center holes. The straight stitch and hinge are the two forms of lacing in most general use. Other methods of splicing fabric and rubber belting are also used in emergencies. Special endless belts with a diagonal splice are supplied by some manufacturers.

A belt should be capable of transmitting from 5 to 25 per cent more power than is actually needed, the excess capacity being governed by the type of drive, the smaller excess for heavy main drives, and the greatest for machine belts. Never use belts the full width of the pulley face, as a slight lack of alingment may cause part of the belt to run beyond the edge of the pulley and perhaps against a shifting finger or pulley flange. This is sure to result in loss of power and perhaps in a badly damaged belt. On the other hand, a belt too narrow necessitates high tension to transmit the required power, thus causing excessive journal friction and the early destruction of the belt.

Generally speaking, single belts, if heavy enough to carry the load, should be used on small pulleys. A single belt should never be wider than 1⅓ times the diameter of the smallest pulley. Where small diameter pulleys and the load would require an unusually wide single belt, it is advisable to substitute narrower pulleys and a narrow double belt. Thin wide belts give the best service; working vertically, a thick narrow vertical belt will not grip the pulley well. Double belts of medium or heavy weight should never be used on pulleys less than 12 inches in diameter, or, even better, on pulleys less than 20 inches in diameter.

Belts too heavy for the load weave back and forth on the pulleys. This is best illustrated by a belt working under intermittent loads, which runs straight while carrying the maximum or propêr load but shows a tendency to weave when the load is reduced.

The tightness with which belts are adjusted to the pulleys is of prime importance. If they are put on too tightly, there is a large unnecessary loss of power from excessive friction at the bearings, to say nothing of the overstrain and injury to the belt itself. If, on the other hand, the belt is too loose, it is likely to flop around and jump from the pulleys, particularly when a load is suddenly thrown on or off. The slacker a belt can be run up to a certain point while doing its work satisfactorily, the greater the economy. In installing belting and taking it up, it should be remembered that certain kinds of belting are affected by weather conditions, lengthening and shortening according to the amount of moisture in the air. Instances have been known where the babbitt was melted out of the boxes and even shafting pulled out of alignment as a result of belting being put on too tight.

Be sure to note carefully whether your shafting is properly in line. More belting is ruined by improperly lined shafting and pulleys than in any other way. The belts under this condition are either kept on the pulleys by guides or rub against hangers and tear or stretch on one edge. Belts so stretched will not run straight and may slip the pulleys at any time. Remember that because your shafting was once in line it does not follow that it will stay so indefinitely. It is a very simple matter to connect two pulleys by a band in such a way that when one pulley is turned the other will go round. It is not at all a simple matter properly to proportion a drive and select the size and quality of belting which will transmit power most effectively and economically.

Shafts should not be located too close together. There should be distance enough between them to allow the belt to recover somewhat from the strain applied on the tight or working side. This distance depends entirely upon the size of the belt and pulleys, but should be sufficient to allow some slight sag to the slack side of the belt. Neither should the shafts be too far apart, for in such cases the weight of the belts draws heavily on the shafting and increases the friction load in the bearings, at the same time causing the belt to swing from side to side and sometimes to run off the pulleys. This constant swing is sometimes the cause of crooked spots in the belt, which wear out rapidly.

It is the best and most economical practice to use pulleys of large diameters, thus obtaining a high belt speed, which means a saving both in the transmission of power and in the wear and tear on the equipment. The speed of the belt being the same, pulleys of large diameters effect a slightly greater transmission of power than do those of small diameter. The speed of the shafts being the same, the advantage in favor of large pulleys over small ones is in proportion to their diameters. Wooden or leather-covered pulleys have a greater transmitting power than iron. Cold-rolled shafting is said to have 30 per cent greater strength than hot rolled. The usual diameters for shafting are from $1\frac{1}{2}$ to $3\frac{1}{2}$ inches. The proper speed is from 300 to 400 revolutions per minute, and its transmitting power is given as $\frac{D^3 \times R}{80} = \text{H. P.}$ D is diameter of shafting in inches, R revolutions per minute.

Shafts that are to be connected with each other by belts should be far enough apart to allow a gentle sag to the belt when in motion. When narrow belts are to be run over small pulleys, 15 feet is a good average, the belt having a sag of from $1\frac{1}{2}$ to 2 inches. For larger belts working on larger pulleys the distance should be from 20 to 25 feet, with a sag of from $2\frac{1}{2}$ to 4 inches. For main belts working on very large pulleys the distance should be from 25 to 30 feet, with a sag of 4 or 5 inches. If the distance between pullies is too great the belt will have an unsteady flapping motion which will tend to destroy both belt and machinery. Regardless of the width of belt or diameter of pulley, the adhesion of the belt to the pulley is the same in all cases, provided the arc of contact and aggregate tension or weight are the same. Thus, other things being equal, a belt will slip just as readily on a pulley 4 feet in diameter as it will on a pulley 2 feet in diameter. To obtain the greatest amount of power from belts the pulleys must be covered with leather. This will allow the belts to run slack and will increase their durability by 25 per cent.

CIRCULAR SAWS.

KINDS AND COSTS.

Circular saws are either of the solid-tooth or inserted-tooth type. The advantages of inserted-tooth saws over solid-tooth saws are: The bits are cheap and can be readily set in position with the special wrench; less experience is required in dressing the saw; there is less filing and gumming; there are fewer saw repairs, which is important in a backwoods locality; and the diameter of the saw remains unchanged during its use. The disadvantages are: The saw kerf is heavy; the teeth are larger and fewer than in a solid-tooth saw; feed is comparatively slow; and the cost is higher.

For big logs and high speed a double circular saw must be used. The two saws are hung to revolve in opposite directions, so that the sawdust from the top saw will not be thrown into the lower one. The advantages of a top saw (double mill) are: It will saw bigger logs than a single mill; it will make a truer cut and saw lumber more evenly; it takes faster feed, saws more lumber, and entails less expense for saws and less repairs. The top saw may remain inactive when small logs are being cut, to avoid using up power. Inserted teeth are not used in a top-saw rig.

The approximate prices of solid-tooth and inserted-tooth saws f. o. b. Seattle, in 1916, were:

Approximate price of solid-tooth and inserted-tooth saws.

Size.	Solid tooth.	Inserted tooth.	Size.	Solid tooth.	Inserted tooth.
40-inch	$23	$44	52-inch	$52	$81
42-inch	26	49	54-inch	58	91
44-inch	30	53	56-inch	66	99
46-inch	36	57	58-inch	74	110
48-inch	41	62	60-inch	83	121
50-inch	47	70	62-inch	94	134

The amount of horsepower required for a circular saw is equal to approximately one-third of the saw's diameter in inches. In large mills each horsepower is supposed to manufacture 1,000 feet of lumber per day; in small mills only one-half that amount.

In ordering a circular saw be careful to give the following specifications in detail:

1. Diameter of the saw in inches.
2. Right or left hand mill.
3. Gauge of saw at center and at rim.
4. Number of teeth in saw.
5. Style or pattern of tooth, solid or inserted.
6. Diameter of mandrel hole, diameter of pinhole and distance, center to center of pinholes.
7. Number of revolutions of saw per minute, while in cut.
8. Greatest feed in inches per revolution and kind of feed.
9. Species of timber to be sawed.
10. Spring or swage set.
11. Horsepower available and size of belt pulley.
12. Engine speed and size of mandrel pulley.

SIZE.

The size of the saw should be governed by the size of the logs to be cut, regardless of the amount of power used. The diameter of the saw should be approximately one and a half times the diameter of the largest log to be cut. For instance, a 36-inch log needs a 54-inch saw, a 40-inch log a 60-inch saw. The width of the widest board which a single circular saw can cut equals the radius of the saw minus 3 inches; that is, a 60-inch circular saw can cut a board or plank 27 inches wide.

HANGING A SAW.

Set the saw plumb and true. Set the saw guide and adjust the guide pins clear of the teeth and just touching the plate. This should be done while the saw is in motion, care being taken that the pins do not push the saw to one side or rub hard enough to cause friction. After screwing the saw up between the collars examine the front or log side of the saw to make sure that it is flat. Never attempt to run a saw that is dishing on the log side, as it will be sure to draw toward the log and be ruined. It does not follow that because one saw works well that another will do so on the same mandrel, or that two saws will hang alike on the same mandrel.

It is absolutely necessary that the saw mandrel should be perfectly level, so that the saw will hang exactly plumb. If it is found to be rounding on the log side, cut a ring of paper about half an inch wide the size of the collar on the outside, oil it, and stick it on the face of the tight collar around the outer cage. Then cut another ring of paper the same width, making the hole the same size as the hole in the saw, put this small ring between the loose collar and the saw, and screw up the collar. If the two rings are not enough, put in more until the saw stands straight and true. If the saw hangs dishing on the log side, reverse the rings of paper; that is, put the small rings between the saw and the fast collar and the large ring against the loose collar. If the saw heats in the center when the mandrel runs cool in the boxes, cool it off and give it a little more lead into the log. If the saw heats in the rim and not in the center, cool it off and give it a little more lead out of the log. Circular saws, revolving as they do at high speed, have a tendency to stretch on the rim, which causes them to cut out of line, dodge knots, run snaky, and become hot in the rim.

HAMMERING A SAW—BLOCKING.

To overcome rim expansion, saws are loosened in the body by hammering. This operation is called "blocking," and the effect on the saw is called tension. If after long use or through accident a saw will not run true and requires hammering, great care should be exercised in bringing it back to proper adjustment. Hammering should hardly ever be done by anyone except an expert. However, if considerable care is exercised, a person who is a fair mechanic should attain good results by practice and observation.

After taking the saw from the mandrel, stand it upright on the floor and examine it carefully on both sides with a straight edge for any inequalities such as lumps. Mark these lumps with chalk on the full or convex side, then place the saw on a block of wood and hammer lightly on the high side. Don't hammer the saw on an anvil, for that will change its tension.

If struck too heavily, the plate will be dented or the lumps knocked through to the other side. When the lumps are knocked too far it is practically impossible to get them back or to restore the equality of the plate. Lumps are sometimes round, but are usually oblong. To remove round lumps, use a round-faced hammer, which leaves an impression one-half inch in diameter. To remove oblong lumps, use a cross-faced hammer. The blows should follow the direction of the lump. Never use an ordinary carpenter's hammer; it will ruin the plate. Do not be content with putting the straight edge on once or twice and hammering a little. After a few blocks, apply the straight edge again, and continue alternatively applying the straight edge and lightly hammering until all inequalities are removed

The saw should then be placed on an anvil (not a wooden block) and examined for tension. Raise one side so that the center of the saw just clears the anvil while the opposite side rests on the bench or wooden board, which should be 1 inch below the surface of the anvil. The body of the saw should be loose enough to drop away from a straight edge placed across the log side evenly from rim to rim. If a 36 inch straight edge is used on a 18 inch saw, you should be able to see light between the straight edge and the center. Always try the long side for this drop. If you find there is not sufficient tension, the plate must be hammered lightly all over on both sides to within 6 or 8 inches of the center and to within 2 inches of the bottom of the sockets. The hammering must be distributed evenly with a round-faced hammer. When adjusting the tension, leave the shanks in the plate.

When hammering a saw it is necessary to know the speed at which it is to run. No saw will run properly unless hammered for the correct speed. A saw when running must be perfectly flat on the log side, although some sawyers prefer to have the edge lean slightly towards the log when the saw is standing still. This means that it is dished a very little on the log side. If, however, any such dish is allowed, it would, of course, be so slight that it could not be distinguished by the naked eye and could hardly be found with a straight edge.

A saw often becomes full on the log side. By this is meant that the rim has been forced away from the log so as to cause the dish on the board side. If there are no uneven spots in the plate other than this dish, the saw can easily be brought back by lightly hammering on the log side.

Always make sure before any hammering is done that the part of the saw which is to receive the blow is well bedded or flat on the anvil. In a perfect saw the apex or point of the V of the socket is central with the plate. If it is not central, the bit, when inserted, will lean to one side. This causes the saw to lead into, or be forced away from the log, as the case may be, and besides dishing the saw, divides the strain on the bit and shank unevenly, causing them to break.

SAW SPEED.

There is a wide difference of opinion as to proper speed of saws. The manufacturers give the maximum speed at which their saws may be operated with safety on the basis of the highest power the saws are calculated to withstand. These speeds can not be used for portable mills because of insufficient power. They are given by saw makers to show what the saw will stand and not what it is supposed to accomplish in practical work every day. While speed is power, it is easy to consume all the power in speed without doing any work. A 48-inch saw run by a 10-horsepower engine should have a speed of 300 revolutions per minute, or to give the best results, 350 revolutions per minute, and should have 24 teeth. The proper saw speed for portable mills running with 20 horsepower and under may be obtained by multiplying the horsepower by 360 and dividing the product by the circumference of the saw.

For 20 horsepower the teeth should be 5 inches apart, which will give 30 teeth to a 48-inch saw. With larger power the teeth should be closer together until they reach the limit of 3 inches apart. As the power be increased the speed of the saw is increased correspondingly.

Saws for cutting hard wood or frozen timber are run at a higher speed than those cutting soft timber, but have more teeth.

If a saw is run at a faster or slower rate than it is hammered for, it is sure to work improperly and give trouble. A saw hammered to run at 500 r. p. m., if run at 350 to 400 r. p. m. in the cut, will be loose in the center and tight around the rim. This will cause it to run snaky or crooked in the log, heat rapidly in the center, and consume a great deal of power. On the other hand, if a saw that is hammered for slow speed, say 300 or 400 r. p. m., is run at 500 r. p. m., it will stretch too much and be tight in the center and loose and wobbly in the rim. This will cause it to heat on the rim, "flutter," run crooked in the log, and consume power uselessly. Improper speed and giving the saw too much lead into the log are the causes of nearly all the trouble experienced with saws in portable mills.

The following table, based on a saw 48 inches in diameter, may be of use in determining proper saw speed.

	Horsepower.					
	6	8	10	12	15	20
Distance between teeth from point to point .... inches..	7	7	6	6	5	5
Teeth in saw .... number..	22	22	24	24	30	30
Speed of saw .... revolutions per minute..	300	300	300	350	400	450

To find the proper speed of larger or smaller saws, multiply the speed given for a 48-inch saw by 48 and divide the product by the size of the saw selected. A larger saw should have a greater number of teeth and a smaller saw a lesser number, the distance apart remaining the same. In the case of a double mill, both saws should have the same rim speed.

FILING.

Solid-tooth saws.—Do not file all the teeth from the same side, especially if each alternate tooth is bent for the set. File the teeth that are bent from you on one side and leave them on a slight bevel, with the outer corners a little the longest, and then reverse the saw and treat the other side in the same manner. Never file any saw to sharp or acute angles at the throat or roots of the teeth, but on circular lines. Any saw is likely to crack from sharp corners. The filing of angles or square corners will cancel the warranty of the saw.

Saw teeth wear narrow at the extreme points; consequently they must be kept spread (swedged) so that they will be widest at the very points. The teeth should be kept as near a uniform shape and distance apart as possible. The back or top of the tooth leads or guides the saw in a measure, and should be filed square across. If a saw is frozen, always thaw it out before attempting to set the teeth.

The greatest wear on the saw is on the under edges of the teeth. File nearly to an edge, but not quite, leaving a short bevel of one-thirty-second of an inch wide on the underside of the point. But in no instance file to a fine point or thin wire edge. Do nearly all the filing on the underside of the teeth and see that they are well spread (swedged) at the points. File square across and have them project alike on both sides of the saw.

On crosscut saws, bevel the front of one tooth one way and the next tooth the opposite way. File rip saws square in front; use a light hammer in swedging, from three-fourths to 1 pound, holding the swedge so that the teeth will spread at the points. Swedge out wide and then dress down to the required width. Most cases of saws "crumbling" are caused by the use of too delicate a swedge. In such cases the corners of the tooth are not strong enough to stand the cutting strain.

Be sure to have a side file for jointing the teeth and keeping all the points the same length on each side of the saw.

The cost of changing solid-tooth saws into inserted-tooth saws is usually $1.50 per tooth plus one-half the list price of a solid-tooth saw of the same size.

Inserted teeth.—It is not economical to spend much time filing inserted teeth, because new points are cheap and are quickly inserted. It should never be necessary to swedge the bits. Light filing on the underside is the only thing required to keep them sharp, because the cutting point tapers gradually and allows plenty of clearance, even though the point is worn almost entirely away. Swedging is likely to injure the shoulders and sockets, unless great care is taken in the operation. Use a light 1-pound hammer and hold the swedge so that the teeth will be spread at the point. Have a cup of oil, together with the teeth, placed conveniently near you at the back of the saw. Take the wrench and place the pins in the holes in the shank, turning the latter so that the hook projects sufficiently to receive the bit. Pick up a tooth with the other hand and dip the grooved segment into the oil, then place it in position and hold it firmly and evenly with the sides of the blade, and at the same time press the wrench downward until the shank fits into its place. The inserted teeth are exact in width and the spread uniformly good. If extra nice work is desired, try a gauge on the side of each tooth. If any are found to project a trifle too far, reduce them with a side file, being careful to preserve the same relief of the corner. No flat surface should be allowed on the side of the teeth, which must be relieved from the very point.

A properly adjusted set of inserted teeth should be used until worn out. Sharpen them as often as necessary by filing on their underside with an inserted tooth file, i. e., a mill file with one round edge. After being sharpened several times they should be relieved on the sides. Should a shank become straight or compressed, because of the saw being run on iron, so that it will not hold the bit firmly, lay it on the anvil and strike it with a hammer on the inner edge until it is expanded sufficiently to hold the bit. Do not try the experiment of bending each alternate tooth for the set. If there is any difficulty in removing an old bit for the purpose of inserting a new one, never hammer the wrench or the bit, but place a blunt cold chisel on the heel of the shank, taking care it does not touch the saw plate, and tap lightly with a hammer. This will start the bit and shank and enable you to remove them easily with the wrench. It is important that all the bits should be of equal length. In order to insure this, get a small segment or section of an old plate that can be easily held in a vise, and insert the bit that needs sharpening or swedging.

For winter sawing use a sharp bevel bit with a narrower cutting point than that used in summer. Saws over 48 inches in diameter, when made thinner than 10 gauge, $\frac{9}{64}$ scant, are not guaranteed.

KEEP YOUR SAWS IN GOOD CONDITION.

If through constant wear a saw becomes weak and limber, the sockets become out of round, and the teeth show a tendency to break or fall out, discard it at once, or send it back to the factory to be fixed up. Saws which have passed through a fire can be repaired if any life is left in the plate, but, of course, can not be made as good as new. Shoulders of inserted-tooth saws that are broken off by running against iron or stone can be welded on again, provided they are not broken off too deep in the plate. If they are broken too deep to weld, the saw can sometimes be renewed by inserting a piece of steel and boring a new socket. The tendency to discard saws or "scrap" them is very general in all mills, big and little, though in many instances, particularly in small mills, the fault lies with the sawyer and not with the saw. A little more attention to the proper care of saws would remedy this evil. If an operator leaves his machinery unhoused, as many operators do, he creates a condition which no amount of instruction can remedy.

LOG DECK.

Every mill should have a good solid log deck or skidway capable of holding from 100 to 200 logs, and built with enough pitch to permit the logs to roll down to the carriage by gravity. Logs should be piled in the yard in such a way that they can be moved to the log deck with the least possible expense and labor. This the small operator commonly fails to do. As often as not the logs are banked in a mud hole. Sometimes 300 or 400 logs are left scattered all about the yard, so that a man and team are kept busy in getting them to the log carriage. When no team is available, the logs are often rolled by hand over slabs, rocks, and bark. The mill is stopped and all hands are called on to help.

This is surely an unnecessary waste of time and energy. Logs should never be dumped in a place from which it will cost at least half as much more to get them to the log deck as it did to load and haul them to the yard. Instead they should be piled on flat skidways in the yard, with ends touching the log road. The hauling team should then be able to keep the saw supplied directly from the woods or from the logs in stock. If the main log deck is kept full all the time the mill is running, there is little chance that logs will litter the yard.

NARROW-GAUGE LUMBER LORRY TRACK.

A narrow-gauge lorry track is almost indispensable to a small mill. The track should be at least 700 feet in length and so constructed that one man can shove a loaded lorry or car over it. The lumber piles should be built on either side of the track, to enable the lorry man to unload the lumber in front of the pile which its grade calls for. Slabs can be moved by the same means. A switch and double track just outside the mill will enable the off-bearer to load an empty car while the lorry man is unloading another in the yard and piling some of the lumber. Lorry trucks have a 26½-inch gauge and 2½-inch tread. The axles are of steel and can be used in wood frames without boxes. The size of the wheels varies from 8 to 16 inches, the weight from 110 to 262 pounds.

From a labor-saving standpoint the lorry track is one of the most important adjuncts to a portable mill outfit. It revolutionizes completely the old manner of handling lumber and does away with the unseemly clutter of slabs, lumber, and other débris around and in the mill.

When water is available a log pond and jack slip are also very desirable.

SETTING UP A PORTABLE MILL.

For foundation timbers place two pieces 10 by 10 inches by 14 feet long on either side of the saw pit (which is supposed to be 3 or 4 feet deep and underneath the husk frame) well bedded in the ground and extending out under the track stringers, taking care to have the one that is under the front end of the husk (where the sawyer stands) placed back far enough to clear the large gear wheel and so that the swinging stirrup will not strike it. One piece 6 by 6 inches by 8 feet, is saddled into the two big pieces spanning the saw pit and underneath the far rail of the track. Never make the mistake of placing the large gear wheel between the pinion and the saw; it should be outside of the pinion. First set the husk frame and secure it to the bed timbers just where you want it. Then put the large gear wheel, with its shaft which carries the rack pin on into its proper bearings, in the central section of track timbers just where it is to work. Then lay this section of track timbers in position on the bed timbers beside the husk frame and push it forward until the large gear comes into proper mesh with the feed pinion on the end of the sliding friction shaft. The inside stringer next to the husk frame has the flat track attached to it, and is higher than the outside stringer, to provide for the difference in height of the flat and V track. The V track is placed on the outside stringer. When track stringers are not shipped with the mill, care should be taken to have the inside stringer higher, in order to bring the top of both tracks on a level. Stringers may be made of any size timber, but where the rack pinion shaft passes through the inside stringer the thickness must not be over 3½ inches, otherwise the track pinion can not be lined with the rack on the front side of the carriage. (When the mill is a large one, No. 4 or over, this does not apply.) When stringers are not over 3½ or 4½ inches the crossties should be placed 18 inches apart, but if heavier stringers are used the ties can be 2 feet or more apart.

The husk and track stringers are not always on the same level on the large cross sills. In most cases the head block base must be elevated by placing a separate piece of timber under each end of the husk on top of the cross sills. Head blocks should be 1 inch above the saw collar. Do not line up under the head blocks, but raise the stringers or lower the husk frame until the blocks are in proper position.

To insure square, accurate lumber the husk frame, track, carriage, and head should be set perfectly level. The nose or front end of the head blocks should clear the saw by one half inch. Set the track perfectly parallel with the front side of the husk frame and see that it is true and level from end to end. After setting the head blocks and set works on the carriage, pull the knees forward until they are even with the front end of the head block bases, then tighten the set-shaft coupling, and after that raise the set shaft so that the pinions will mesh deep enough in the racks under the knees to take up all lost motion. Then fasten the set-shaft bearings on the side of the head block base. These bearings are adjustable, to provide for raising and lowering the set shaft. Run the knees back and forth a few times to see if they come up even with the nose of the bases. If there is any variation it can be adjusted by the set-shaft coupling. The mandrel must be level, and the journal must fill the boxes and should be so constructed that there will be no movement endwise when the mandrel is running. The mandrel must fill the eye of the saw, but enter freely. Always use a wrench and never a hammer on nut of arbor. The pins in the collar must have a fair bearing. Sometimes when driving them in a burr is thrown up. If not carefully filed off this tends to throw the saw over to one side.

After the mill is set up and before starting, move the carriage slowly by means of the sliding paper friction. When everything is in good order this friction can be turned easily by hand and the carriage moved in either direction. If it can not be easily moved, examine the gear and pinion to see that they do not mesh too deep or bind against the flanges; see that the rack pinions do not mesh too deeply in the rack under the carriage; also examine the set collars on the track axles to see whether they are too tight and need slight loosening. The carriage will run hard when the track is not level and straight. Give the saw from one-eighth to one-fourth inch lead in 20 feet, according to the condition of the saw and character of the lumber to be made. More or less lead is given the saw by means of set screws on either side of the main mandrel bearing. Do not try to lead the saw by pulling it over with the guide. The wood pins in a saw guide clear the bottom of the saw teeth by an inch.

A good way to give proper lead to a saw is to turn the rear head block up opposite the center of the saw, and fasten a stick or board on the head block so that the end of it is set one-eighth inch from the saw. Then run the carriage back until this stick is 20 feet from the center of the saw. Then stretch a line from the end of the stick along the face of the saw, so that it touches the saw on both edges. If it does not touch the saw on both edges, adjust the main mandrel box by the set screws on each side of it until it does. You then have one-eighth inch lead in 20 feet. This is called slewing the mandrel to regulate the lead. The same result may be obtained by sighting over the saw and fixing the saw plane for a radius of 10 feet. This may be done by placing two shafts vertically into the ground 10 feet from the saw center, behind and in front of the saw. Then a horizontal stick is fastened to a head block so as to just touch the forward staff. Then the carriage is gigged backward to the other vertical staff, where the horizontal stick must lack exactly one-eighth of an inch from touching. The belt holes or set screws in the boxes of most mandrels are slotted, and it is only necessary to loosen the set screws and move the boxes one way or the other by a few light blows of a hammer in order to get the proper lead.

OPERATING THE MILL.

First see that everything has been set up according to directions and all nuts are tight. Then with kerosene or benzine carefully clean the turned surfaces of the friction disk and the sliding friction shaft, taking care that no oil or grease or any foreign substances get on the paper frictions. Nothing should ever be put on these; simply keep them clean and dry, and covered so that the greasy sawdust will not get into them and cut them out. If the frictions should become wet, a little fine sawdust will dry them nicely. All bearings must be kept clean and supplied with an abundance of good machine oil. Where grease cups are used, they must be filled with a good grade of grease and kept covered so that dirt and grit will not work into them. It is a good plan to go over all of these at regular intervals and clean out the oil holes.

Start the mill up slowly, see that the bearings are not too tight, and also that they do not run hot. Do not try to cut up a log the very first thing. Run the mill empty for a while and watch every working part to see that there is no heating and that nothing needs adjustment. Be a little slack on the frictions until you learn just how to work them nicely. By carelessness or crowding the pressure on the frictions before everything is working nice and easy, you may grind a flat place on the sliding paper friction and destroy its usefulness. When everything is properly set up and correctly adjusted a very slight pressure of the hand on the feed lever is all that is necessary to run the carriage when sawing.

Do not fasten the stirrup in any way at the bottom, but leave it free to swing back and forth as the feed lever is pressed to or from the disk. The upper end of the stirrup must be fastened to the iron lug at the bottom of the yoke of the sliding box with a cap screw. It must also fit up in the socket of the sliding box, so that it will move the box forward and bring the sliding friction up against the disk when the feed lever is pressed forward.

SAWING.

For your first sawing pick out moderate-sized logs. Do not try to get full capacity out of your mill for the first day or two. See that the log is dogged firmly to the knees, saw slowly and get the "hang" of your mill, and let the working parts all get in good running order. Use an extra quantity of oil until everything is working easily and smoothly. By studying the set works on the carriage, you will soon learn to cut any desired thickness of lumber.

A working knowledge of how to run a saw can be easily acquired. Of course, it is the better way, if possible, to take lessons from some one who has already had experience of this kind. If, however, no one is available, start very slowly until you know just what feed you can give the various-sized logs. To insure that the logs will be cut to best advantage, the amateur sawmill operator should take a pair of dividers or a piece of string and a pencil and lay out several circles representing sizes of different logs, and then draw in the ends of the boards or stock in the way you wish to cut them. You will soon learn the best way to turn and handle logs so as to get the most desirable and best-sized pieces out of each log. Keep your mill clean and take away the sawdust and do not let everything become covered up with oil and dirt. Keep all nuts tightened up and look your mill over every day before starting. You will thus save many shutdowns.

As opposed to this plan, many sawyers cut their logs up "alive" without turning the log at all, edging the boards afterwards with the circular saw. This system may be dismissed as the very "limit" in wasteful milling. Another plan, which is followed by quite a number of operators, is to slab off the boards on four sides of the log, leaving the wane on each board, and cutting up the squared portion into planks or boards. This method of sawing is made to act as a grading medium for the lumber; and, accordingly, we find such operators selling their lumber as square edge, sheeting, and waney edge, with square-edge lumber topping the price list and the others in order. In reality, the waney edge, if run through an edger, is the best lumber cut from the outside of the log, although by this method classed as lowest in price and grade. However, this is a little better than the "saw-'em-alive" system, for it enables the operator to secure a higher price for his product, although mixing his grades unintentionally. The better and more successful class of operators keep close supervision on the sawyer to see that he gets the maximum amount of good lumber out of each log, grade their lumber more or less carefully, and pile it in the yard according to grade and dimensions. They are generally equipped with a light planer and edger, and some few have a floorer. A few of these operators occasionally buy the better quality of lumber from the less thrifty millmen of the "saw-'em-alive" class, grade, edge, and surface it, and make more profit than the man who logged and milled it in the first instance.

The greatest defect in connection with a portable mill seems to lie in the irregular size of the lumber produced, arising from the inability of the sawyer to get the proper "lead" in sawing. Next to this is lack of sufficient power in boiler and engine. The effective horsepower available in most cases is far below what the mill owner supposes. He may have an engine of 25 or 30 indicated horsepower, while the effective power for working purposes may be only 10, owing to the excessive friction, badly lined machinery, poor belting, and leaking cylinders. Under such conditions the only thing for the millman to do is to employ a competent mechanic and have his mill overhauled and put in proper shape. The operator who does not keep his mill in good working order had better get rid of it at once.

PILING LUMBER.

Every pile of lumber should rest on three strong horizontal ground sills. The front sill should be higher than the middle and back sills. The front of the pile should be given an overhanging "batter" to protect it from the rain. The usual pitch is 1 foot to 10. The tiers of boards are kept apart by three or four well-seasoned crosspieces called "sticks" sawed about 1 inch square and placed directly over one another in the tiers. The usual width of a pile is from 6 to 10 feet, and the distance between piles ought to be 3 feet. Each pile must contain equal lengths, as "overlaps" are sure to get spoiled. A sufficient air chamber should be left in the center of the pile so that the stock may dry rapidly. When fresh-sawed lumber is allowed to touch, discoloration of the portions touching is sure to ensue. Each pile should have a roof 12 inches high in front and 6 inches high in the back, projecting on all sides. In order to prevent end cracks, the sticking should be placed exactly at the ends and slightly projecting over them. Inch boards and planks should be handled carefully when being piled or loaded on wagons. If slammed down violently, as is the custom, end checks are likely to develop into cracks or splits which spoil the grade of the piece.

A 2-acre yard or piling space is necessary for a small mill carrying 50,000 board feet in stock. Select good level ground for your lumber yard and see that the lumber is neatly piled according to size and grade. You will get more for your lumber if you can show it to a customer in this condition than where it is all dumped in a heap. Remember that the mill run is ordinarily sold at a loss to the millman, as the buyer grades it according to the lowest quality in the whole pile and not according to the high grades which, in properly piled stock, would be sold separately and at higher prices.

FIRE PROTECTION.

There is little use in going to the expense of putting up a mill if it is allowed to burn down. Fire is an ever present danger. For writing a policy on the ordinary small mill fire insurance companies charge a premium equivalent to 20 per cent of the mill's value, which makes the expense of insurance prohibitive in most cases. It is up to the owner, then, to provide fire protection. This can be done by installing liquid chemical extinguishers or dry chemical extinguishers or water hose that can be attached to the boiler. Liquid chemical extinguishers may freeze in winter, and dry chemical extinguishers are expensive. The most practical equipment for a portable mill is water hose, and in the following table is given the capacity of such hose at various pressures:

Capacity of water hose at various pressures (size of nozzle, 1 inch).

	Pressure of nozzle.			
	40 pounds.	60 pounds.	80 pounds.	100 pounds.
Pressure at pump with 100 feet of 2½-inch rubber hose, pounds	48	73	97	121
Gallons per minute	155	189	219	245
Horizontal distance thrown, feet	109	142	168	186
Vertical distance, feet	79	108	131	148

LOGGING.

SAW CREW AND EQUIPMENT.

For a mill cutting 10,000 feet (from 9 to 11 logs per thousand), two pony gangs are necessary. If the sawyers do not understand felling and how to get straight logs and good lengths out of the tree after it is felled, it may be necessary to have an experienced under-cutter with the two gangs. One of the crew should be able to file and fit a saw properly. Incidentally, the saw should be a good one, so that the filer does not waste time trying to sharpen it, or the sawyers lose 5 or 10 minutes in every cut. The same applies to axes. An axe is intended to chop, not to bruise timber, and must be kept sharp if it is to be used effectively. It should be ground on a grindstone, not rubbed with a file. It would be interesting to know how many men working round a small mill understand how to grind an ax properly. The knowledge of making ax handles, oxbows, and yokes is a lost art, yet when a man is working 20 or 30 miles away from a hardware store and there are no ax handles in camp, this knowledge would come in handy.

A saw gang of two men, equipped with the proper tools and working in timber which will run from 8 to 10 logs per 1,000 feet, will ordinarily cut between 5,000 and 6,000 feet per day, the logs varying in length from 12 to 16 feet. The average wage for this class of work in 1916 was approximately $3 per day, board not included. If the average cut is 5,000 feet per day, the cost per 1,000 for sawing would be $1.20. On the other hand, take two equally good sawyers and equip them with a rusty worn-out saw, poor axes, and no wedges, and the best they can cut will be around 3,000 feet per day. At the same rate of wages this timber will cost the operator $2 per 1,000 for bucking alone. The latter crew loses the price of a new crosscut saw every seven working days, or the price of a new ax every day. Lack of good cant hooks, swamp hooks, logging chains, etc., cuts further into the profit margin.

Each saw gang should be provided with a pair of good steel wedges (preferably forged crucible tool steel) and a maul to drive them. The usual practice is for the sawyer to stop and whittle out a wooden wedge when he needs it. In doing this a gang of sawyers loses more time each day than is represented by the cost of a set of good steel wedges. If not equipped with a maul, a sawyer will spoil a dollar ax in five minutes trying to drive steel wedges with it. Tying the wedges together with baling wire when not in use will prevent them from being lost. Each gang should have a flat bottle, easily carried in the hip pocket, to hold kerosene with which to spray the saw occasionally.

CROSSCUT SAWS.

Ordinarily a 6-foot saw is best adapted to the size of timber in the inter-mountain country. In some stands, however, a longer saw is necessary. The back of the saw is always somewhat thinner than the gauge of the teeth, and usually about two gauges heavier on the ends, which makes it stiffer and prevents "buckling" when the sawyer pushes a trifle on the back stroke.

There are two kinds of teeth, viz., the cutting teeth and rakers. Only the points of the cutters actually cut into the fiber; the rakers are meant to plane off the fiber severed by the cutting teeth and to drag the sawdust out of the cut. A curved saw will cut faster and with less exertion than a straight saw.

A filer's outfit consists of a jointer, a raker tooth gauge, files, a raker swage, a set block, and a light hammer. In fitting a saw the teeth should first be jointed or made uniform in length. To do this place a 9-inch file in the jointer and, by means of the large screw, spring it to suit the curve of the saw and pass it lightly over the points of the teeth until it touches the shortest tooth. Then place the tooth gauge over the cleaner drag teeth and file them down to the gauge, afterwards filing them squarely across to a point as desired.

If the saw requires setting, place the setting block on a log that has been leveled to receive it, then place the saw upon the setting block so that the point of the tooth to be set projects over the apex of the beveled surface fully one-fourth of an inch. Then give two or three strokes with a light hammer on the side of the tooth over the apex fully one-fourth inch from the point which will usually give the required set. Then move the tooth forward over the bevel or channelling block and strike it a sharp blow directly over the point. This will make a slight depression below the point of the tooth, which serves to keep the cutting edges flush, giving them greater cutting power and relieving the pressure on the back of the tooth. Regulate the set by gauging each tooth with the tooth gauge. Take the gauge in the left hand and place it against the side of the saw. The point on the short end indicates the least set, and the point on the long end indicates a little more set. Experienced saw filers say that setting should never go lower than half the length of the tooth, should never exceed twice the gauge of the tooth, and that more set is needed in long saws and for softwoods than for short saws and hardwoods.

The side file should be used to remove any feather edge or burr left in filing and to even up the set perfectly. This can be done while the saw is in the vise or filing clamp. Place an 8-inch mill bastard file in the recess on the side file and tighten the screws holding the file. Be particular that the saw is held firmly in the clamp. Pass the center of the file lightly against the teeth until the burr is removed and the set evened up. Care should be taken not to use the side file more than is necessary to remove the feather edge and even up the set. The wrench may be used to lessen the set if necessary, but should never be used to set the teeth. This should be done only with a hammer upon the setting block or on any hard and slightly beveled surface. Notice carefully the manner in which the saw is filed when new, and file as nearly as possible along the same lines.

To secure the best results, the saw should be sharpened after being set. If it is desired merely to swage the point of the rakers, this may be done with a light blow of the hammer on the extreme point of the raker teeth. A raker gauge, the flange of which rests on the points of the cutting teeth, is useful for inexperienced saw filers. The filer can reduce the projecting points with the file until stopped by the edge of the gauge. In this manner tooth after tooth can be rapidly and correctly reduced to an even length by any unskilled operator.

A saw now on the market has the teeth in groups of three, the center one a cutting tooth and the ones on either side rakers. The cutting teeth are set, but the rakers are not. The saw can be used for any kind of timber, the teeth varying in size according to the work they are to do. The raker teeth are beveled to form a sharp point and have cutting edges, thus forming a combined raker and cutting tooth. This is the distinctive feature of the saw.

In ripping, the bottom of the kerf is cut sideways. This causes the saw to rip faster and easier than the old-style ripsaw, which works chisel fashion and has to cut the wood almost square across the grain. The set which is given the cutting teeth assists the rakers very materially by cutting the kerf a little wider than the raker teeth and by cutting the sides of the kerf perfectly smooth. This causes the saw to run freely and easily. Ripping is done by cutting, instead of tearing, as is the case with the old-style ripsaw.

In crosscutting, the teeth cut in four different lines in the kerf. In this way the fibers are cut in small, short pieces and are much easier to break out than if they extended clear across the kerf. As all teeth are of the same length they loosen the sawdust clear to the bottom of the kerf, thus removing much friction that occurs where a square raker tooth is used. This style of tooth works equally well in ripping, miter, or crosscutting, and in all kinds of soft and hard wood, as well as where knots occur and the grain is tangled in many different directions. That the saw will cut faster and run easier either way than any other saw that is filed for one purpose is claimed by the makers. No gauge is required to cut the raker teeth down to the proper length, nor does it materially interfere with the proper working of the saw if raker or cutting teeth are filed shorter than the others here and there.

In filing this saw, joint down the points of the teeth with a flat file until the file touches every tooth. Set the cutting teeth (the middle teeth in every group) a good ways back, making a gradual bend, then pass a flat file along the sides of the teeth to even the set. File all the raker and cutting teeth almost to a point; then file to form a chisel point. Pass a flat file along the sides of the teeth to remove the featheredge. Keep the saw in a leaning position, as it enables you to file a full bevel. The more bevel you give the teeth the faster the saw will cut. Carefully notice the shape of the teeth in a new saw and file to conform with them as nearly as possible.

The file should be laid between the teeth, with its flat side on the bevel of the tooth you are filing, but without touching the point of the next tooth. Each tooth is filed separately. The teeth, when new, are cut deeper than is required for the angle of the tooth, so that the bevel is full only a little ways down from the point. After a few filings, however, the bevel will be full clear down to the base of the tooth. In this way every filing will gum all teeth except the big gullet, and that should be filed frequently to afford plenty of room for cleaning out the sawdust. This can be done with the edge of a flat file, or with a round one, if you prefer a round gullet. Do not file the gullet wider than it is in a new saw.

NOTCHING AND FELLING.

The proper method of felling, bucking, skidding, and loading logs is one of the principal subjects upon which the average mill operator needs information. When not properly handled, it is likely to be one of the costliest items in his business.

Run a cut in the butt of the tree to be felled as deep as the size of the tree calls for and at right angles to the direction in which the tree is supposed to fall. Take the axe and chip out a notch above the cut or below it, leaving a clear notch in the butt. When the bark is rough at the base of the tree, chop a girdle round in the line of the saw cut. But before doing this, select the best ground on which to fall the tree, to prevent breaking or pinching when the tree is being made into logs. If the tree to be felled leans heavily in one direction, it is the better plan to notch a little to one side or the other to which the tree leans, and after starting the cut in the opposite side of the tree (notch) "saw round" or cut the corners, leaving a couple of inches of the fiber between the saw cut and the notch uncut on one corner. In this way the strain which will be exerted when the tree is falling will pull the body of the tree in the direction you want it to fall. "Sawing round" prevents a leaning tree from splitting on the stump and possibly breaking the saw.

In felling very tall straight timber with a heavy crown it is sometimes difficult to ascertain in which direction the tree leans the heaviest until the saw is half way in the cut, when the saw becomes fast and requires heavy wedging to loosen it. In wedging such trees, great care should be taken to prevent them from coming back over the cut and injuring the men. Thick wooden wedges are used in extreme cases of this kind. Men felling timber should always yell "timber" before the tree starts to fall, and when it starts, should remove the saw and stand clear.

The tree should never be completely severed from the stump. Ordinarily a tree, when falling, will break off 2 or 3 inches of the fiber between the saw cut and the notch. A heavy notch which leaves a long stub on the butt log should be avoided.

Sawyers should endeavor to fall the timber as favorably as possible for skidding purposes, and in a way to prevent the cuts from pinching the saw. A little care along those lines will perceptibly affect the volume of cut. Keep count of the number of logs each gang cuts per day.

If a tree lodges in another when falling, fall the one it is lodged against at once, or, if you are an expert in judging the direction of falling timber, fall another one against the one lodged so as to bring down the latter. Never continue working around a lodged tree. It is likely to hit you when you are not expecting it.

After a tree is down, the limbs should be cut off close to the bole and thrown out of the way. The next step is to measure off each log by placing the measuring stick (8 feet long) flat on the bole and clipping a piece of bark off at the end of the first 8 feet, then moving the stick forward to the exact length of the log to be cut and making a plain notch in the bark for 2 or 3 inches over the exact length, to allow for trimming and broomage on the ends. Walk alongside the tree, and not on it, when measuring the logs.

All logs should be straight. If a sweep occurs in a tree, cut in the sweep. If a cut pinches, cut a double length log and get the teamster to pull it out to where it can be sawed without "pinching." If there is a break, cut up to the break and start your next log on the other side of it. If a crotch is in the tree, cut out the crotch, but don't waste any lumber in doing it. If a felled tree balances on a knoll or on another tree, prop the light end. If part of the bole is limby and part is smooth, have as much of the smooth portion as possible in one or two logs and the rough knotty portions in different cuts. If a tree is punky or dozy or has dry knots or blind punk, measure off the logs so as to confine the defective part, as much as possible, to one log length. Sever each log completely from the one next to it. Cut up all down timber and skid it before felling any green timber on top of it. Do not leave a "chip" at the bottom of each cut—it annoys the skidder. Do not saw anything but the log. If the ground is higher than the bottom of the log, "saw short" or dig a channel for the saw; look out for rocks. Do not allow your saw to "run"—make a square cut every time.

Where there is an undercutter on the job—and there should not be one unless he understands his business—it is a good plan to have him notch a number of trees ahead for the sawyers to fell. This enables the sawyers to work to better advantage with the skidders. Before bucking up any of the down timber, the undercutter can trim the trees ahead of the saw gang, who will then have no brush to retard their work, while the skidding teamster can figure ahead on his skid roads. Brush disposal should follow skidding.

A logger should cut his log lengths with a strict regard to the current prices for lumber of different lengths. Then, if he grades his lumber, he will not sell what should be 2-inch No. 1 shop, 12 inches wide, 18 and 20 feet long, as No. 3 common.

PREPARING SPECIAL PRODUCTS.

Mine props, telegraph, telephone, or power poles should be trimmed and peeled, piled carefully according to lengths, and air dried before shipment. The successful marketing of such special products depends largely on how they are prepared. If a mine operator gives an order for a carload or two of 7-foot mine props and gets an assortment of green props of irregular lengths with the ends chopped or "chewed" off, the bark on, and snags all over the bole, be assured that if he accepts them at all it will be at a reduced price and that he will look to some other source for material in the future. Careless preparation has in more than one instance prevented the utilization of large quantities of timber which, if properly handled, would have been entirely acceptable to the mining operators. When a mine operator has to "fit" each prop after receiving it, that is, trim the knots off, peel it, and square the ends to the proper length, he is doing work which you should have done, and he knows it. Again, if a mine operator places an order for green lodgepole pine, do not try to pass off dead timber of the same species or ship a mixed carload of alpine fir, white fir, aspen, and lodgepole. You may work off one carload, but that will be all.

When such material has to be shipped by rail, freight charges become an important factor in favor of the local man. Yet the foreign shipper who prepares his mine timbers properly can, on a 400-mile rail haul, create and hold a market for the same species of mine prop material which the local operator can not sell at all, although his product is 300 or more miles closer to the consuming center. The sole reason is that the local man does not prepare his material properly. Tell the mine operator just what species of timber you have for sale and ship it to him properly made up and you will hold his trade.

THE MAIN LOGGING ROAD.

A logger should never forget that it is cheaper and more economical to haul a load of logs a quarter of a mile to his mill on a downhill road or on the level than it is to haul a similar load a rod uphill. Run the main road through the stand to be cut, if the ground is level, or below the timber if the stand is on a side hill. Keep your cutting area compact, and when you make a skid road, take out all the timber tributary to it before you move to the next one.

If you have to haul up a hill, double up; do not figure on loads that one team can handle over a hill. If the road is too steep, use a block and line. Do not waste time in trying to force your single team to haul a load which a four-horse team can hardly get away with. If the ground is rough and rocky, use dynamite and build a good main road and keep it in good order. Breakage and wear and tear on equipment and horses over a bad road will buy a lot of powder and keep a road in fine shape.

One thousand feet of logs weigh approximately 4½ tons. It would take a good team to haul such a load over a paved street. Why try to haul it over rocks, ruts, and logs when logging in the hills with a pair of cayuses? A logger should remember that the value of the time lost and the injury to teams and equipment while struggling to pull off impossible stunts in logging is always a prolific source of loss. If iced or even snow roads were possible. it might be different; but a dirt road must be well built.

If the timber is large, use a gin pole and crotch line in loading; use bunk chains or corner binds in holding the first tier of logs on the truck or sled. A couple of logs on top of the binding chain will hold them taut. A spring pole is dangerous to use and not a safe bind on a load of logs. Have your toggle bind and grabs on the unloading side.

It sometimes happens that part of the timber is relatively close to the mill and part farther away and not so accessible, and the character of the road such as to prevent the hauling of a maximum load every trip. In this case it will be found economical to skid up the "close" timber along the main road and "top off" the long-haul loads to their full capacity on their way to the mill.

SKIDDING.

Skid or dray roads are not supposed to receive as much attention in construction as main logging roads. Nevertheless, they should be kept reasonably clear and should be laid out with considerable care. On steep hillsides it is sometimes feasible to build bench roads along the face of the hill. The logs are skidded to the nearest bench and then rolled by hand to the next bench, and so on down to the base of the hill, where the trucks can handle them to the landing. In some cases where the power (oxen or horses) is available and the distance to the mill is not over half a mile logging "from the stump" can be done to advantage.

Use a dray or go-devil to skid the logs out to the main road. Skidding logs by chain, except on a downhill haul, or on level, smooth ground, is waste of time. A team can haul three logs on a dray over rough ground with less effort than it takes to haul one log by tongs or a chain.

When a tree is cut into logs in the woods, the skidding teamster should place his dray alongside the end log with a short skid resting on the dray, set his swamp hook low down on the far side of the log, hitch his whiffletrees to the swamp-hook chain, start up his team, and roll the log onto the bunk of the dray. If the log is a big one, it should be peeled on the side which drags on the ground. If your team can haul more than one log, fasten the first log on the dray and move up to the next and repeat the same operation. If the tree has fallen in a position difficult to load easily, hitch on the block and line and boost the log out to where it can be handled. Do not make your team pull a log on the chains or tongs out of a pothole or from between two rocks or stumps simply for the satisfaction of seeing them do it. Get the log out in the easiest and quickest way possible and spare the team and rigging.

When loading logs on the skidway, place the ends which will come on the front of the load even. This can be done very easily when filling the skidway, but not so easily with the skidway full. "Rossing," or peeling the bark off logs, particularly large ones, will make skidding easier and prevent insect attack. "Snipping" the forward end is also resorted to, but a log so large and heavy that the front end digs into the ground when being hauled on a chain or tongs should be loaded on a dray.

Logs on a travois or dray should be balanced on the bunk of the dray in such a manner that approximately two-thirds of the weight of the load is behind the bunk. Logs will haul easier when the heaviest end drags on the ground. They should be so balanced that if the runners should strike on a rock, the dray will partially lift and slide over the obstruction. If the load rests wholly on the bunk and runners, the team will be unable to lift the load.

In addition to the main binding chain, on the end of which there is usually a round hook, a skidding teamster should always be provided with a grab or fit hook with which to keep his load snug and to take up slack when necessary.

Snags and stumps should be removed or cut low along a skid road. Windfalls and loose rocks should be thrown out of the road, and chuck holes filled up. Never try to haul or skid logs over a swamp or bog unless it is frozen hard enough to carry teams; build a corduroy road. If water is available, the skidding teams should be fed in the timber. Build a brush shelter equipped with feed boxes and keep a few days' supply of hay and grain stored there. Remove the harness while the horses are feeding. Never, if it can be avoided, throw grain on the ground for teams to eat. Have fly nets for the horses in warm weather and keep the horses shod.

Skidding equipment consists of good harness, whiffletrees and trace chains, logging chain, cold shuts, tongs, swamp hook, cant hooks, ax, "Sampson," travois or dray, skids, block and line. A jackknife dray, with a 3-foot bunk spiked and fitted with bunk chains, is preferred. This dray is so constructed that if it strikes any obstruction it will "jackknife," or partially fold. There should be spike skids for decking on the skidways when teams are not available. Both smooth and spiked skids should be shod with iron on one end to prevent slipping. When teams are not available and the logs are small (10 per 1,000 feet), two men can load ("spike") logs onto a truck or "deck"; but when large logs are being handled a gin pole and loading line (parbuckle) is necessary. If the men are inexperienced in the use of cant hooks, as is usually the case in small operations, a crotch loading line is preferable, because one end of the log can not slide ahead when rolling up the skid, as it is likely to do in a single line, particularly if the log is larger at one end than at the other. Several small logs can be "sent up" in a crotch line at one trip, but not so easily in a single line.

SKIDWAYS.

A skidway should be centrally located and the approach so graded or so gradual as to avoid an uphill haul. The end of the skids should be blocked or preferably bridged apart, so that the team will have firm footing when crossing the skids. The front end or head blocks should, if possible, be high enough to make it possible, when the skids rest on them, to roll on the first tier of logs by hand. The skids should have a slight slope toward the front. The ground alongside the skidway should be cleared of brush, so that the longest log will not catch on the ends. It is not worth while to build a skidway at all if logs can not be handled from it faster and with less effort than they can be loaded among the brush on the dray road. The purpose of a skidway is to expedite the work. A poor skidway will retard it.

Chaining or tonging logs onto a skidway, except on a close haul, is not recommended. A tonged or chained log, unless turned on the chain or tongs before it strikes the skids, will tear up the approach and make hard hauling. A dray is much better, for besides hauling more logs to a trip it is not so hard on the approach to the skids.

A skidway should be 4 feet wide or wider, if the logs are double length, and long enough to hold at least one day's haul without decking. The skids should be tolerably straight, of equal height, and free from knots. If the skids are unusually long they should be blocked up in the center to prevent them from sagging or breaking. If possible, it is a good plan to embed the hind end of the skids in the ground. If the logs are small and light, a light-built skidway will answer the purpose; if heavy logs are to be loaded, make it solid.

I have seen four men with two teams struggle for two hours to load an 800 board-foot log on a truck from smooth skids and fail, while all the time the teams stood idle. Yet with an improvised parbuckle (a rope spliced to a logging chain) the same log was loaded with the aid of a team in five minutes. In this instance the haul to the mill was a quarter of a mile, and it took four men and two teams one-half a day to land that log on the mill yard, and cost the operator $7, or at the rate of $9 per 1,000 for loading and hauling alone. The log was the butt cut of a five-log tree and not entirely severed from the next log. Instead of skidding the five logs to the main road where there was a good skidway chance, and where that log and another one could have been loaded by hand upon the truck, the operator attempted to load the log at the stump. The heaviest part of the haul was from the stump to the main road. Owing to the position of the logs, only one could be loaded on the truck, even though the four horses might have handled more than one. The men had no cant hook or swamp hook with which to roll the log over and no loading line or parbuckle by which to load it. Handspikes and an old peevy and a piece of a logging chain were the only equipment available. The operator, on being asked how he expected to load logs with such equipment, replied: "Oh, all our logs are not as big as this one; the boys usually breast them up." If this man had had a dray or go-devil, a swamp hook, cant hook, and loading line, with a skidway close by on the main road, his teamster could have loaded this log on the dray, hauled it to the skidway with the other logs in the tree, and loaded two logs on the truck very easily. The hauling team could have made six trips a day instead of two, and the loading and hauling would have cost $1.80 per 1,000 instead of $9.

In practice it invariably happens that a small-mill operator who watches this part of the work carefully makes a success of his business. The average operator usually lets the job by contract to inexperienced men at twice what it could be, or ought to be, done for, in order to be rid of it. At a conservative estimate the logging to small mills costs $1.50 to $2.50 per 1,000 more than it ought to cost. This unnecessary expense swallows up the greater portion of the profits that can be made in operating a small mill. Misdirected energy in this line means a loss to the operator which can not be made up.

CHUTES.

When the country to be logged is split up into narrow canyons or gulches and is not too steep, chute logging can be carried on with more or less success. When the chute has a heavy grade and the logs attain a high velocity at the landing, the loss in breakage is too heavy to justify this method of handling timber. In fact, chute logging is most successful where the grade is so light that horses are necessary to keep the logs moving. Twenty or more logs dogged in a string propelled by horsepower can be taken over a chute rather cheaply.

If the distance to "chute" exceeds a mile, beats should be laid off in one-fourth mile sections, with a man and team to keep each one clear. In small operations, of course, one man and team or one horse and man are enough. In addition, a team or single horse is needed to keep clear the point where the logs are being delivered to the chute.

When the grade is heavy, soft steel goosenecks are used to check the speed of the logs, and the delivery end of the chute is elevated in order to insure that the logs will fall flat instead of striking "end on." Logs handled in this way are generally split in the end and badly bruised up, even when they are landed in water.

Chutes are usually built of 20-foot logs laid two abreast and hewed trough-shaped by cutting away the inner faces. The logs are drift bolted to bed pieces that are sunk firmly in the ground, 12 to 14 feet apart. The average size of chute timbers is 14 inches at the top. The average cost of chute construction is $3 per rod.

A chute in which logs run by gravity is called a running chute, and one with which horses are used, a trailing chute. Sometimes logs will run on one part of a chute and are trailed on other parts. Heavy logs will run on lighter grades than small, light, short logs. Trailing chutes should be greased, or, if the weather permits, iced.

The main trouble about chutes is that very often they are not constructed properly, particularly chutes built around sharp curves. A running chute should never be built with a bad curve. It costs too much to keep in repair, and logs will be constantly jumping. In small jobs when the ground is adapted to chutes and the operator has some experience in chute building, their use is strongly recommended. The cost will not be more than that of a good road, which few operators build. Moreover, the cost of trying to get his logs to his mill over a bad road will be avoided. Small operators seldom cut logs over 18 feet in length, and these can be taken around a pretty sharp curve on a trailing chute. The operator must, however, build a good chute, or else he can not use it.

Big logs should be peeled or rossed for chuting. The average cost of trailing logs in a chute varies from 50 cents to $1.50 per 1,000. Well-built chutes will, if kept in repair, last from 7 to 10 years. If not needed for that period the timbers can be taken up, and after all the spikes are removed can be cut into lumber. The usual grades for long logs are—

	Per cent.
Dry chute	15 to 20
Ice chutes	4 to 8
Water chutes	3 to 6

For short logs, ties, mine props, and field posts the grades are—

	Per cent.
Dry chute	25 to 35
Iced chutes	8 to 12
Water chutes	5 to 8

Hardwoods are better than conifers for chute construction. The grade of inlet must be steep, and the outlet in running chutes should be into a pond of water of sufficient depth to keep the logs from striking bottom.

LOADING LOGS.

It is important to load in the right way, for every time a log is moved unnecessarily there is a cut in the operator's profit.

Dig a hole for the gin pole deep enough to prevent the pole from sliding out but loose enough to admit of the pole being pushed back a few inches in order to clear the logs when the load is being hauled away. Make the gin pole fast to the sway bars of the truck or sled with a short chain and grab. See that the truck or sled is in proper line with the front end of the logs to be loaded. Set up the skids with the ironed tips resting on the bunks or sway bars and (as this is addressed to portable millmen) use a crotch line with light swamp hooks on either end. Make each dog fast in the end of the bunk or in the side of the sway bar. After it is passed round the log hitch the loading line on the whiffletree or equalizer, start the team up slowly, and see that the log is coming fairly on to the bunks and not faster on one end than the other. Be careful that one end does not miss the bunk and fall between them on the reach or hounds.

A crotch line is recommended for portable mill loggers because it is much safer for inexperienced loaders, although slower than a single line. After a little practice in the use of cant hooks, however, a green man will soon learn to "cut" a log which is inclined to go endwise on the skids (cannon). In such cases the loader should step in front of the lower end of the log and catch it smartly with his hook, at the same time pulling so that the end will slide forward on the skids. If he happens to be on the side which is moving too fast, he should catch the log smartly from behind and by a sudden pull hold that end while the other or slower end slides forward. This operation requires quickness and dexterity in the use of the hook. A green hand is likely to catch his hook too deep in the timber, with the result that he gets a severe jolt and his cant-hook stock is broken.

A loader should never stand behind a log going up in the chain. When a log is safely on the load he should have the next log rolled down to the skids, help to pull back the chain far enough to encircle the log and to reach the load, and hand the swamp hook to the loader, who makes it fast to a log already on the load, usually the log in front of the one beside which he intends the log in the chain to lie. The top loader grabs the log with his cant hook, after fastening his swamp hook in the log to prevent it rolling over when the team pulls on the chain, yells to the teamster to go ahead, and keeps clear of the log as it rolls into its place. A good top loader figures ahead on where he intends to place his logs in order that the load will balance. If one side of a sled load of logs is heavier than the other, the load usually becomes "bunk bound" and will not steer properly, and naturally, if a truck is not loaded properly, it will tip over.

Corner binds and binding chains should be used in hauling logs on a truck or sled. Binding or spring poles are not reliable, besides being dangerous if they happen to slip. Good stout block brakes are a requisite on steep hills. These should be secure enough to enable a logger to stop "dead" on the steepest hill—rough locking the hind wheels with stout chains if necessary. The neck yoke, breast straps, and hame straps should be carefully looked over every day and kept in proper shape.

The cross haul for the team should be brushed out and the ground leveled off so that the team can haul to good advantage and turn around easily. Sometimes when a block is used on the gin pole the loading team pulls parallel to the load and along the road. If this can be done, the crosshaul can be dispensed with.

When a load is taken off the front of a skidway the remaining logs should be rolled down to the front, in order to make room for more logs on the end. It does not pay to deck (pile up) logs in a small operation. Medium-sized logs can be decked a couple of tiers high in spiked skids quite easily, thereby saving the expense of a decking team and teamster.

If you are hauling on a snow or ice road, keep the space in front of the skidways on which the sled stands free from bark and chips. Figure on loading the heaviest logs on the bottom of the tier. If the sawyers leave any knots on the logs, trim them off close to the bole and see that the binding chains or toggles are grabbed on the unloading side. If the sawyers have left a "chip" on the bottom of the cut, set the "Samson" (which should be higher than the log) on the near side, drop the swamp hook over the far side of the log, and fasten in the dog. Make the swamp-hook chain or logging chain fast with a half hitch around the top of the "Samson," and hitch the team to the chain. One pull from the team usually breaks the chip or turns the log over so that the teamster can reach it with his ax. If the teamster is not equipped with a swamp hook and does not understand how to use a "Samson," he can take a "roll" on the log with his log chain, with the draft down low on the side of the log opposite his team, and if the log is not too heavy and the chip too large, can lift the log far enough with his team to enable him to cut the chip with his ax and roll the log on to the dray.

The jackknife dray has a roller in front set on gudgeons in the nose of the runners. A stout bunk armed with spikes connects the runners. If the dray strikes any obstruction it will "jackknife" or partially fold, hence the name.

SCALING.

There are many rules for the measurement of saw logs. None of them, however, is mathematically exact, simply because a saw log is not a mathematical figure. Also, the unit of sawed lumber is 144 square inches, or a piece of board 12 inches square and 1 inch thick. Any rule purporting to give the actual amount of board feet which a log of a given diameter will saw out is, therefore, only an approximation.

The Decimal C rule is the one adopted by the Forest Service for scaling Government timber. This rule is made by dropping the unit figures of the Scribner rule and accepting the nearest tens as correct values. The contents of logs as given by the Scribner rule were found by drawing diagrams outlining the smaller end sections of logs and computing the board feet in the boards that could be sawed from them.

The Doyle rule, which is extensively used, is based on the following formula: Deduct 4 inches from the diameter of the log, square one-fourth of the remainder, and multiply by the length of the log in feet. This gives the approximate contents in board feet. Another rule of more ancient origin, which gives the same results, is to deduct 4 inches from the diameter of the log as an allowance for sawdust and slabs, then multiply the remainder by half itself, and the product by the length of the log in feet, and divide by 8. The quotient will be the number of square feet contained in the log. Still another giving the same contents is, deduct 4 inches and square the remainder, adding to or deducting for the aliquot parts of a longer or shorter log.

The Doyle rule gives less values in board feet for 16-foot logs up to 28 inches in diameter than the Scribner or the Decimal C rule, and above that diameter gives greater values. Between 22 and 28 inches in diameter, however, there is only a slight difference between the two rules. The 14 single logs ranging from 8 to 21 inches in diameter scale by the latter rule 12 per cent more than with the Doyle, though logs scaled by the Decimal C rule overrun the log scale by an average of 10 per cent when manufactured into lumber.

The requirements of the Forest Service call for the measurement of sound material in the log irrespective of grade. Allowance is made for bad defects, such as rot, shake, check, cat face, crooks, wormholes, etc., or a serious combination of one or more such defects in one log. In timber of high commercial value, logs are classed as cull that have two-thirds of the gross scale defective; in the case of inferior species, such as fir, lodgepole, and balsam, one-half of the gross scale.

It very often happens that timber or logs containing only one-third of the gross scale in merchantable lumber will not pay the cost of handling unless the lumber obtained is high grade. In such cases the scaler should possess a knowledge of grades and be qualified to differentiate between a log that pays for merchandizing and one that does not.

The only true analysis of what a defective log will scale is found by seeing it opened up in the mill. A scaler's ability to make proper allowance for defect hinges on his experience in this particular respect. Defects such as large loose or unsound knots, an occasional knot hole, a great deal of pitch and pockets, some red rot and bad season checks, large wormholes, and any amount of blue stain, unless combined in one log, affect merely the quality or grade and not the quantity of merchantable lumber produced from the log. They are therefore not recognized as justifying a scaler in discounting the gross scale of a log.

The fact of logs overrunning the log scale—that is, cutting more board feet than is shown on the Decimal C rule—should not be considered by a scaler in making allowance for defect. This matter of overrun is, in Forest Service sales, taken into account when appraising the timber and establishing the stumpage price, and is not relevant to scaling.

The following grading rules for western yellow pine logs are suggested by the Forest Service for use in eastern Oregon and Washington, and could be profitably used by portable mill operators in any section of the country:

No. 1 clear logs shall be 22 inches or over in diameter inside the bark at the small end and not less than 10 feet long. They shall be reasonably straight grained, practically surface clear, and with not less than 25 per cent of their scaled contents capable of being cut into C select and better.

No. 2 shop logs shall be 18 inches and over in diameter inside the bark at the small end, not less than 8 feet long, and with not less than 30 per cent of their scaled contents capable of being cut into No. 2 shop or better.

No. 3 rough logs shall be 6 inches and over in diameter inside the bark at the small end and not less than 8 feet long, and having defects which unfit them for classification in the two above grades.

C select grade means lumber 4 inches wide or wider. Knots, blue stain, some pitch or season checks are admissible. A 4-inch or 6-inch piece would show light traces of pitch or season checks. Heavier indications of pitch and season checks are permissible in wider boards, but should not be scattered all over the board. Medium blue stain, covering one-third of the face, if not in combination with other defects, is admissible.

No. 2 shop means a grade of lumber valued for cutting purposes only. The grade of No. 2 door cuttings will admit of one defect in one piece. This may be a small sound knot not to exceed five-eighths of an inch in diameter, or a light-blue stain which does not extend over more than one-half the surface of the piece on one side, or, in the absence of all other defects, one small season check not to exceed 8 inches in length and showing on one side of the piece only. Each plank of No. 2 shop should contain either one of the following: At least 25 per cent of No. 1 door cuttings or not less than 33⅓ per cent of No. 1 and No. 2 door cuttings combined.

Rough logs of the grade lower than Nos. 1 and 2 consist of lumber the general appearance of which is coarse, admitting many defects in inch lumber and all defects common to dimension lumber that do not materially impair the strength of the piece.

LOGGING OUTFIT

Following is a logging outfit for class A mills and its cost:

4 teams and harness, at $400	$1,600
Saws, axes, wedges, cant hooks, chains, and blacksmith outfit	150
2 logging trucks	250
2 lumber wagons	200
Camp cook outfit	100
Sleeping camp	50
Lubricating oil and kerosene	20
Total	2,370

The following equipment would be suitable for a class B mill:

5 teams and harness, at $400	$2,000
Saws, axes, wedges, cant hooks, chains, and blacksmith outfit	150
2 logging trucks	250
2 lumber wagons	200
2 drays	10
Cook camp outfit	150
Sleeping camp	70
Lubricating oil and kerosene	20
Total	2,850

Below is given an outfit of blacksmith tools for small loggers and others doing their own work:

Coal and borax.
60-pound anvil, with hardies.
Leg vice, 35-pound, opens 4 inches.
Lever forge, 60-pound.
Upright drill.
Round shank drill.
6 taps and 3 set dies.
Cold chisel.
Assorted punches.
Tongs, pincers, and rasps.
Farrier's knife.
Shoeing and turning hammer.
Hoof parer, horseshoes and nails, calks, and sling for shoeing unruly horses.

The following constitute logging accessories:

Logging trucks, Montana, $130 each.
Lindsay self-loading skidder, 2 wheels, 24 inches diameter; 5-inch tire, $30; 8-inch tire, $35; 10-inch tire, $40.
Four-wheel logging trucks: Weight, 1,200 pounds, capacity, 4,000 pounds;[1] weight, 1,800 pounds, capacity, 10,000 pounds.
Large-size cant hooks, 5½-foot stock.
Swamp hooks, steel.
Skidding tongs, opening 24 to 30 inches.
Emery or carborundum wheels, 10 inches diameter and 1 inch thick.
Runners, with McLaren's casting, finished and ironed, set of four.
Bolster plates, four to a sled.
Gudgeon pins.
King bolt.
Pole caps.
Tote sleds.
Logging sleds (no bunks), complete. (Shipping weight, 2,400 pounds.)
Big wheels, 7 to 10 feet high.
Lumber buggies, capacity, 2,000 pounds; weight, 300 pounds.
Dump carts, 2-horse, weight, 1,900 pounds; 1-horse, weight, 1,300 pounds.
Two-horse wagons with box, weight, 1,350 pounds.
Boot calks for driving boots.
Square shovels, D handles, 38 inches over all.

[1] This rig includes whiffletrees, evener, stay chains, neck yoke, or tongue chains. Brake, $10 extra. Two-wheel carts with straight or arched axle.

Round shovels, D handles, 38 inches over all.
Shovels, long-handled, square pointed, 61 inches over all; round pointed, 61 inches over all.
Spades, D-handles, 39 inches over all.
Snow shovels.
Earth augers for holes 5 to 8 inches diameter; 8 to 14 inches diameter.
Posthole tamping bars.
Hayforks, 4 tines.
Manure forks.
Manure D-handled forks.
Railroad and grading plows.
Wheel scrapers.
Slushers.
Slushers, double bottom.
Wheelbarrows, tubular steel frames.
Wheelbarrows, wood.
Dynamite, 40 per cent (20 cents per pound in 25 to 50 pound boxes).
Caps, per 1,000, No. 3.
Safety fuse, 50 feet in coil, 2 coils in a package, per 1,000 feet, single tape.
Blasting powder, 25-pound sheet-iron kegs, black F.
Blasting machine, size 13 by 9 by $6\frac{3}{4}$ inches, weight 35 pounds, capacity 1 to 30 holes.

Following is a list of tools and stores:

Pickarooms or hookaroons: Axe-handle shape; straight handle.
Pike poles with straight pikes, 14 to 16 feet.
Pike poles with hook and straight pike.
Straight pike only.
Pike with hook.
Adz handles.
Spike maul handles.
Riveting and farrier's handles.
Round chain hooks.
Corner bind or fid hooks, flat.
Half round fid hooks.
Boom chains made to order.
Log jammer blocks.
Loading blocks.
Skidding tongs.
Rollway breaking hooks.
Loading hooks.
Swamp hooks, steel.
Cant hooks, iron clasp.
Cant hooks, steel clasp.
Peavies, malleable socket.
Peavies, steel socket.
Cant hook and peavy handles.
Cant hook and peavy hooks.
Top loader, heavy steel.
Peavy sockets.
Cant-hook clasps.
Toe rings, iron.
Toe rings, steel.
Hook and peavy bolts.

Ax handles, single bitted.
Ax handles, double bitted.
Axes, single bitted.
Axes, double bitted.
Broadax handles, Michigan.
Pick and mattock handles.
Maul handles.
Broadaxes.
Driving batteaux.
Paddles.
Anchors.
Road rutters.
Badger road rutter.
Snowplow and road rutter (Eau Claire).
Logging sled stock: Pole, rock elm; white oak. Finished and iron pole, and roll, elm; roll, oak.

APPENDIX.

WATER.

One gallon of water weighs 8$\frac{1}{3}$ pounds and contains 231 cubic inches. One cubic foot of water contains 7$\frac{1}{2}$ gallons, 1,728 cubic inches, and weighs 62$\frac{1}{2}$ pounds. One cubic inch of water evaporated under atmospheric pressure is converted into approximately one cubic foot of steam. The height of a column of fresh water equal to 1 pound pressure per square inch is 2.31 feet. In usual computation this is taken at 2 feet.

To find the pressure in pounds per square inch of a column of water, multiply the height of the column in feet by 0.434. Approximately every foot elevation is equal to one-half pound pressure per square inch, allowing for friction.

Pressure of water for each foot in height.

Feet in height.	Pounds per square inch.	Feet in height.	Pounds per square inch.	Feet in height.	Pounds per square inch.
1	0.43	15	6.49	50	21.65
2	.86	20	8.66	70	30.32
5	2.16	25	10.82	80	34.65
10	4.33	40	17.32	100	43.31

To find the velocity in feet per minute necessary to discharge a given volume of water in a given time, multiply the number of cubic feet of water by 144 and divide the product by the area of the pipe in inches. Doubling the diameter of a pipe increases its capacity four times. A miner's inch of water is approximately equal to a supply of 12 gallons a minute.

Measurement of an open stream by velocity and cross section.—Measure the depth of the water at from 6 to 12 points across the stream at equal distances between. Add all the depths in feet and inches together and divide by the number of measurements. This will give the average depth, which when multiplied by the width will give the area of the cross section.

The velocity of the stream can be found by laying off 100 feet along the bank and throwing a float into the middle, noting the time taken in passing over 100 feet. Do this a number of times and take the average. Then divide this distance by the time, and it will be the velocity at the surface. As the top of the stream flows faster than the bottom or sides the average velocity being approximately 83 per cent of the surface velocity at the middle, it is recommended to measure a distance of 120 feet for the float to run and reckon it as 100.

Multiply the area of the cross section by the velocity of the stream in feet per minute, and the result will be within 12 per cent of the discharge in cubic feet per minute.

To compute the capacity of pumping engines.—Multiply the area of the water piston in inches by the distance it travels in inches in a given time. Deduct 3 per cent for slip and rod displacement. The product divided by 231 gives the number of gallons in the time named.

Barometric pressure varies at different altitudes with an equivalent variation in the vertical suction lift of the pumps.

An "atmosphere," $14\frac{7}{10}$ pounds, at 62° F. is equal to a column of mercury 29.9 inches high, or a column of water $33\frac{9}{10}$ feet high.

Water at high temperature can not be raised any considerable distance by suction, as the vapor discharged from the water follows the receding piston of the pump and resists the entrance of the water; consequently, to pump hot water always have the supply above the pump so that it will be supplied from a head.

STEAM.

Under the ordinary atmospheric pressure of 14.7 pounds per square inch, water boils at 212° F., passing off as steam, the temperature at which it boils varying with the variation in the pressure. Steam rising from water at its boiling point (212° F.) has a pressure equal to the atmosphere, viz, 14.7 pounds to the square inch at sea level. To evaporate 1 cubic foot of water requires the consumption of $7\frac{1}{2}$ pounds ordinary coal, or about 1 pound of coal or $2\frac{1}{2}$ pounds wood to 1 gallon of water. Steam is transparent, colorless, dry, and invisible except when partly condensed, when the mist makes it visible. Dry steam does not contain any free moisture; wet steam contains free moisture in the form of spray or mist and has the same temperature as dry saturated steam of the same pressure. Saturated steam is steam in its normal state, that is, steam whose temperature is that due its pressure, by which is meant steam at the same temperature as that of the water from which it was generated and upon which it rests. Superheated steam is steam at a temperature above that due its pressure after leaving the liquid from which it is generated.

One cubic inch of water evaporated into steam at 212° F. becomes 1,645 cubic inches, that is, nearly 1 cubic foot. Water introduced into superheated steam will be vaporized until the steam becomes saturated and its temperature becomes that due its pressure. Cold water or water at a lower temperature than that of the steam, when introduced into saturated steam, will condense some of it, thus lowering both the temperature and pressure of the rest until the temperature again equals that due its pressure. The speed of steam for designers of steam piping is about a mile and one-half per minute and sometimes much more.

There is a saving of approximately 1 per cent in fuel burned for every 11° that feed water is warmed before being run into the boiler. If sufficient exhaust steam is available and cold water at 70° is raised to 210° the saving in fuel will approximate 12 per cent.

FUEL.

One cord of air-dried hickory or hard maple weighs about 4,500 pounds and is equal as fuel to 2,000 pounds coal.

One cord of air-dried white oak weighs about 3,850 pounds and is equal to 1,715 pounds coal.

One cord of air-dried red oak or black oak weighs 3,250 pounds and is equal to 1,450 pounds coal.

One cord air-dried poplar, chestnut, or elm weighs 2,350 pounds and is equal to 1,050 pounds coal.

One cord of air-dried yellow, white, or lodgepole pine weighs 2,000 pounds and is equal to 625 pounds coal.

From the foregoing data it is safe to assume that $2\frac{1}{4}$ pounds dry wood is equal to 1 pound average-quality soft coal, and that the fuel value of the same weight of different woods is very nearly the same. That is, 1 pound of hickory is worth no more for fuel than a pound of pine, assuming both to be dry. It is important that the wood be dry, as each 10 per cent of water or moisture will detract 12 per cent from its value as fuel. This fact should be noted by small mill operators who use green slabs for fuel and sell the dry ones. In one case observed by the writer the dry slabs were sold in the mill yard at 1 cent each and the green ones used to fire with. During the afternoon the mill stopped four times, averaging 15 minutes each time, in order to get sufficient steam to run the saw, and then the sawyer had to "back up" the carriage several times to enable the saw to gather sufficient speed to cut a foot or two farther into the log.

Evaporative power of 1 pound of various fuels at atmospheric pressure (14.7 pounds).—1 pound good coal will evaporate 10 pounds water; 1 pound crude petroleum will evaporate 16 pounds water; 1 pound natural gas (25 cubic feet) will evaporate 20 pounds water.

WATER WHEELS.

The horse power of falling water is expressed in the formula $\frac{V \times H \times 62.5}{33,000}$, which means that the volume of discharge in cubic feet per minute multiplied by the height of fall in feet, multiplied by the weight of a cubic foot of water, and divided by 33,000, gives the horse power per minute.

Amount of water required to develop a given horsepower with a given available effective head.[1]

Effective head.	Flow of water per minute.			
	10 H. P.	20 H. P.	30 H. P.	40 H. P.
	Cubic feet.	*Cubic feet.*	*Cubic feet.*	*Cubic feet.*
50 feet	125	250	375	500
60 feet	104	208	312	416
70 feet	88	177	266	355
80 feet	77	155	232	311
90 feet	70	140	210	280
100 feet	63	125	186	248

[1] Horsepower based In 85 per cent efficiency of the wheel.

Water wheels are built either overshot, breast, undershot, or turbines.

Overshot wheel.—The effective power is 60 to 75 per cent of possible power. The proper velocity of the circumference is 5 feet per second and is equal to approximately one-half the velocity of the water. The water velocity must be greater than the rim of the wheel. In falls of from 20 to 40 feet in height the overshot wheel is more effective than a turbine. The buckets should have a capacity three times as large as the volume of water actually carried, should have holes in the bottom in order to allow the escape of air, and have a depth of from 12 to 14 inches and be 12 inches apart at least from centers. The speed necessary to run a sawmill with this power is obtained by countershaft or gears.

Breast wheel.—The water acts by weight and impact, dropping vertically into the buckets. The efficiency varies from 45 to 65 per cent of possible power and works best in falls from 8 to 16 feet having a discharge from 20 to 80 cubic feet per second. The speed of the wheel should be such as to fill the buckets one-half their capacity each revolution. Other forms of breast wheel are called "high breast," "low breast," and "flutter" wheels. The old form of water wheel had radial buckets, but the modern idea is in favor of curved buckets which give a higher efficiency.

Undershot wheel.—The undershot or current wheels have a low efficiency and are usually anchored in rapid streams in such a position as to be safe during floods or high water. The size of the wheel varies. It usually has 12 blades, each one of which is submerged as it passes directly under the axle. All of those wheels have become practically obsolete in this country.

The Pelton turbine is a type of axial flow impulse turbine in which a small jet issues from a nozzle and strikes on a series of cups of the shape of two hemispheres joined together at the center by a straight thin rib. By this means the jet is split and returned without serious shock. The speed of the rim should be one-half of the jet velocity to give the highest efficiency. The water works by impact and pressure and not by weight, as in vertical wheels, and greater speed is directly transmitted, which makes a turbine wheel better adapted for sawmill purposes than a vertical one.

The advantage of water-power mills is less operating expense. The disadvantages are: Water-power mills are not portable; the source of power not uniform or continuous; not as capable of control as steam; and in the majority of cases results in low speed and low capacity.

ENGINE FRICTION BRAKE.

The engine friction brake is an instrument for measuring the power which an engine can give off for external use. It is usually applied to a flywheel and will give results if the belting is properly adjusted and the shafting in line. The object of a flywheel is to compensate for the irregular turning movements, and to prevent the consequent variation of speed from exceeding certain predetermined limits. The energy stored in a flywheel varies as the square of the velocity and directly as the weight. In the case of solid cast iron of good quality the velocity of the rim should not exceed 80 feet per second and in built-up wheels 50 feet.

STANDARD HORSEPOWER.

Standard horsepower is 33,000 foot-pounds. A foot-pound is 1 pound lifted 1 foot per minute or any equivalent amount of force, such as one-half pound lifted 2 feet or 12 pounds lifted 1 inch per minute. The horsepower of an engine may be determined by the following formula: H. P.$=\frac{2 x_3 PLAN}{3\ ,000}$. When P=mean effective pressure, L=length of the stroke in feet, A=average net area of piston, and N=number of revolutions per minute. To calculate horsepower, multiply the area of the piston in square inches by the speed of the piston in feet per minute and divide the product by 33,000. The result is the horsepower value of 1 pound mean effective pressure, which, if multiplied by the whole mean effective pressure, will give the indicated horsepower. The net effective horsepower is the indicated horsepower less the friction of the engine.

GEARS.

The word "diameter" when applied to gears is always understood to mean the pitch diameters. Diameter pitch of the gear is the number of teeth to each inch of its pitch diameter. If a gear has 40 teeth and the pitch diameter is 4 inches, the diametral pitch is 10, or, properly speaking, the gear is 10 diametral pitch. Circular pitch is the distance from the center of one tooth to the center of the next tooth measured along the pitch circle. If this distance is one-half inch, the gear is said to be one-half circular pitch.

CALCULATING THE SPEED OF SAWS, PULLEYS, AND DRUMS.

EXAMPLES.

1. A 30-inch pulley, making 180 revolutions per minute, drives a countershaft with a 12-inch pulley. What is the speed of the smaller pulley? $180 \times 30 \div 12 = 450$ r. p. m.

2. A countershaft is to make 450 r. p. m., driven by a 30-inch pulley making 180 r. p. m. What will be the diameter of the countershaft pulley? $180 \times 30 \div 450 = 12$-inch pulley.

3. What will be the diameter of a pulley making 180 r. p. m. to drive a 12-inch pulley 450 r. p. m.? $450 \times 12 \div 180 = 30$-inch pulley.

SAW GAUGE.

Stubs's standard English gauge is in general use by saw manufacturers in this country.

No. of wire gauge.	Fractional equivalents.	Decimal equivalent.
		Inch.
No. 1	$\frac{5}{16}$ inch scant	.300
No. 2	$\frac{9}{32}$ inch full	.284
No. 3	$\frac{17}{64}$ inch scant	.259
No. 4	$\frac{15}{64}$ inch full	.238
No. 5	$\frac{7}{32}$ inch full	.220
No. 6	$\frac{13}{64}$ inch scant	.203
No. 7	$\frac{11}{64}$ inch full	.180
No. 8	$\frac{11}{64}$ inch scant	.165
No. 9	$\frac{9}{64}$ inch full	.148
No. 10	$\frac{9}{64}$ inch scant	.134
No. 11	$\frac{1}{8}$ inch scant	.120
No. 12	$\frac{7}{64}$ inch scant	.109
No. 13	$\frac{3}{32}$ inch full	.095
No. 14	$\frac{5}{64}$ inch full	.083
No. 15	$\frac{5}{64}$ scant	.072
No. 16	$\frac{1}{16}$ inch full	.065
No. 17	$\frac{1}{16}$ inch scant	.058
No. 18	$\frac{3}{64}$ inch full	.049

MILL-MACHINERY TERMS.

Back (of a saw tooth)—The upper or convex part of a saw tooth. The lower or concave portion is called the face.

Bit—A tooth used in an inserted tooth saw. The knives used on the cutter heads of surfacing machines to cut a tongue and groove.

Board mill—One that makes a specialty of 1-inch and 2-inch lumber as compared to a timber mill which cuts lumber of greater thickness.

Bumper—A device placed at each end of the carriage run to absorb the shock. It is usually a piston fitted with a cylinder which contains live steam or air, buffer.

Cant flipper—Two or more horizontal bars placed in a line of live rolls; the outer ends are fastened to a common shaft attached to the piston of a steam cylinder and the near ends are pivoted to a firm support. On elevating the outer ends, cants and boards are transferred to a temporary storage point behind the gang resaw or edger.

Carriage—A frame on which are mounted the head blocks, set work and other mechanism for holding the log while it is being sawed and also for advancing the log toward the saw line after a cut has been made. The carriage frame is mounted on trucks which travel on tracks. The carriage is moved by steam, feed cable, or rack and pinion.

Carriage dog—A steel tooth, several of which are attached to a carriage knee and operated by a lever. Their object is to hold the log firmly on the carriage.

Carriage feed—The power used to drive the carriage back and forth. It may consist of a rack and pinion, a cable device, or a large cylinder and piston. In large mills the steam cylinder is used for short carriages and the cable for long ones. In portable mills the rack and pinion or cable feed is used.

Carriage receder—A device on the under side of a carriage which automatically shunts the carriage frame on its axles about $\frac{5}{8}$ inch away from the saw line. The carriage receder is used only in band mills; its use prevents the log from hitting the band saw.

Chip breaker—A roller or bar in front of cutter knives on a planing machine to prevent splinters from being torn from the face of the board as it is being surfaced.

Circular gang mill—A machine usually used to cut 1-inch flooring strips from 4 inch and 6 inch cants.

Circular saw—One having cutting teeth on the circumference of the plate.

Dished saw—Circular saws may assume a shape like that of a dish. The concave side is spoken of as the dished side.

Double-cutting band saw—Has teeth on both edges and cuts on both forward and backward run.

Double mill—A mill having two head saws.

Edger—A frame supporting an arbor on which are mounted several saws, feed rolls, press rolls, and power transmission gear. It is used to square-edge lumber and also to rip it.

Edge stacker—A machine which piles lumber on edge on dry-kiln trucks.

Feed rolls—Live rollers with a smooth, corrugated, or rough surface which holds the lumber and pushes it into an edger, resaw, planer, etc.

Felloe—One of the segments of the rim of a wheel between the spokes and the tire.

Floorer—A planer and matcher combined which makes flooring.

Friction nigger—A long lever armed with teeth used to turn logs on a carriage.

Gang edger—An edger that has fixed saws.

Gang mill—A machine with a heavy frame supporting a sash which carries straight saw blades. The sash runs in upright slides and is driven from below

Gang saw—One from 6 to 10 inches wide and 4 feet in length with teeth on one edge, suspended in the sash of the gang and cutting on the down strokes. Syn.—Gate saw.

Gauge—The thickness of a saw blade.

Head block—The part of a carriage which holds the log and upon which it rests. Each head block consists of a base, a knee, a taper set, and a rack and pinion gear.

High track or dollyway—An elevated tramway which runs from the sawmill to the yard.

Hog—A refuse grinder.

Hookaroon, pickaroon—A curved pike fitted to handle, used in pulling ties or lumber into place.

Horizontal band saw—A band saw which runs horizontally.

Husk—The frame supporting the arbor and other parts of a circular saw.

Inserted tooth circular saw—One in which removable shanks and bits are inserted in the sockets on the rim.

Jump saw—One that can be raised or lowered in a vertical line.

Jack slip—The trough up which the bull chain hauls the logs.

Knee—The part of a carriage holding the dogs and also the levers operating both the dogs and the taper set.

Log deck—The platform in a sawmill upon which logs are stored preparatory to placing them on the carriage.

Log lift—Cable slings, spaced several feet apart—employed to lift logs from water.

Loose—A saw is said to be loose when the surface falls away too much from the straight edge.

Lumber buggy—Dolly lumber truck.

Lumber jack—A tripod armed with a blunt spike on top, used as a fulcrum to pass the lumber up to the lumber piler.

Matcher—A surfacing machine used in a planing mill for finishing lumber of average width and thickness. Syn.—Joiner.

Out of round—A circular saw is said to be out of round when it is not a perfect circle.

Overhead trimmer—One which has the saws hung above the table.

Pond saw—A power-driven drag saw used to cut logs in a mill pond.

Press roll—A live roll which holds the lumber against the feed roll when passing through a machine.

Resaw—A circular or band mill used to resaw boards, cants, plank, timbers. Syn.—Pony band mill slab saw.

Rift gang mill—A machine for cutting edge-grained flooring strips from a cant. It consists of a number of small circulars set on the arbor of an edger.

Rock saw—A circular saw or a planer head which removes a wide kerf on the upper surface of the log in front of the cut of the head saw.

Rotary veneer machine—A machine that cuts or peels a thin endless sheet of wood from a round log.

Sash saw—An upright band of steel toothed on one edge stretched in a sash or frame and used singly usually in a water-power mill of limited capacity.

Saw arbor—The shaft and bearings on which a circular saw is mounted.

Saw guide—A device for steadying a circular or band saw.

Screw rollers—Rollers with a coarse thread which throw the board or slab to one side as the piece passes over it.

Set beam—A shaft on a sawmill carriage connected with the set works bearing pinions, one of which meshes into a rack in each headblock and moves the knees back or forth as desired.

Setting block—A small steel block on which the tooth of a crosscut saw is placed and then struck with a hammer to give it the proper set.

Set works—The mechanism on a sawmill carriage by means of which the setter advances the knees and the log toward the saw line after a piece has been cut from the log.

Set-works scale—A dial on a sawmill carriage which shows the distance between the saw line and the face of the knee.

Shank—Device for locking inserted teeth into the sockets of a circular saw.

Shotgun feed—Steam feed.

Single band—A band saw with one cutting edge; a double band is toothed on both edges.

Single mill—A mill with one head saw.

Sizer—A machine for surfacing timber.

Slasher—Several circulars mounted on the same line from 16 to 24 inches apart for cutting up slabs, edging, etc.

Solid-tooth circular saw—One in which the teeth are cut into the rim of the saw.

Spring set—When one tooth in a saw is sprung slightly to the right and the next one to the left alternately; crosscut and narrow band saws are spring set.

Standard band mill—One having a 50,000 daily capacity from a single band.

Steam feed (or shotgun feed)—A long cylinder with a piston which is fixed to the rear end of the carriage and propels it back and forth.

Steam niggers—A heavy-toothed lever worked by steam cylinders which is used to turn logs on the carriage.

Sticker—Small pieces of boards placed between courses in a lumber pile or a machine used in a sash, door, and blind factory for shaping doors, sash rails, sash bars, and muntins.

Swage—A tool used to spread the points of the teeth of a saw.

Swage set—Hammering the points of the teeth to a width greater than the thickness of the saw. Head saws are usually spring set and some rip saws also.

Taper lever—A lever attached to the knee of a carriage headblock which throws either knee out of alignment when cutting churn-butted logs.

Tension—To make a band or circular looser in the middle than on the cutting edge, by hammering.

Throat—The rounded cavity below the points in which sawdust gathers and is carried from the cut.

Tire—Is that part of a band-saw blade 1 inch or more back from the throats which has not been stretched to conform with the segment to which the rest of the blade is tensioned. This leaves the saw tighter at the tire than it is in the middle. The width of the tire varies with the width of the saw and the amount of tension carried.

To gig a carriage—Running the carriage back after a board is cut from the log.

To gum a saw—To grind out the throats of a saw.

To hammer a saw—To round it with a hammer in order to adjust the tension.

To hang a saw—To place a saw in position ready for operation.

To jack logs—To pull logs from the pond into the mill on an endless spiked chain. Syn.—Bull chain, jacker, log haul.

Top saw—The upper of two circular saws on a head saw, both being on the same husk.

Whip saw—A saw operated by two men used to cut logs into lumber. Syn.—Pit saw.

LUMBER TERMS.

Backing board—The last board in the log left on the carriage.

Barn boards—Boards used for barn siding. The cracks between the boards are covered with battens.

Base—Interior trim which is fastened to the walls of a room at the floor line.

Battens—Narrow lumber used to cover cracks between siding boards.

Bevel cribbing—Boards beveled on both edges, used to cover the sides of a corncrib.

Bevel siding—Lap siding, siding weather board.

Block setter—One who operates the set works on a sawmill carriage.
Blued lumber—Lumber, the sapwood of which has been stained by fungi.
Box boards—Lumber of a specific quality from which boxes are manufactured.
Box shooks—Pieces of lumber cut to size for boxes but not assembled.
Break down—To cut a log into cants or of a size which can be sawed on the main saw.
Bull head—A term used by sawmill filer to describe the action of a saw when it leads in or out of a cut.
Cant—A log which has been slabbed on one or more sides.
Casehardened lumber—Material, the exterior of which is dry, while the interior remains moist. The result of quick drying in an overheated kiln.
Ceiling—Lumber usually finished on one side only and used for wainscoting, ceiling rooms, etc.
Chimney—An opening left from top to bottom in a lumber pile to admit air and hasten drying.
Chipped grain—A defect in lumber caused by the grain of the wood being torn out in patches by the action of the planer knives.
Clapboard—Siding 4 to 6 inches wide and 4 to 20 feet long tapering to a thin edge on one side.
Case knot—One surrounded wholly or partially by pitch or bark.
Coarse-grain lumber—Material with wide annual rings.
Comb grained—The best quality of quarter-sawed lumber, the growth rings of which are nearly at right angles to the face of the board.
Common boards—Applied to four grades of lumber of a quality inferior to finish. The widths run from 4 to 12 inches.
Common dimensions—Applied to 2-inch stock ranging from 4 to 12 inches wide and 3-inch stock from 6 to 12 inches wide.
Custom sawing—The sawing of lumber under contract prices per 1,000 board feet.
Carriage setter—Rides on the front end of the carriage and sets the dogs which hold the log in place.
Drop siding—A pattern of lumber used to cover the exterior sides of buildings. Syn.—Cove siding, German siding, patent siding, rustic.
Face side—That side of a board which shows the best quality.
Featheredge—When a board is found thinner on one edge than it is on the other it is said to have a featheredge. Term also used to describe an over-sharpened cutting edge.
Feed—The length of lumber cut at one revolution of the saw, expressed in inches.
Fencing—A grade of rough inch lumber 4 to 6 inches wide.
Fine grain—Lumber having the annual rings close together.
Finish lumber—The higher grade of lumber used for interior finish in buildings.
Five-ply veneer—Made up of five pieces of veneer glued one to the other, also called laminated wood.
Flitch—A thick piece of lumber with wane on the edge.
Furring—A narrow strip of inch lumber which is nailed to rafters and joints as a backing for laths.
Jointed flooring—A flooring strip which instead of being tongued and grooved has the sides cut on a bevel edge. Syn.—Lap siding.
Joist—A dimension timber used to support the floor of a building.
Knocked down—A machine or article taken apart in order to facilitate shipping. Abbreviated as K. D.
Large knot—One that is over 1½ inches diameter.

Load—In foreign markets 1,680 pounds weight of lumber or 50 cubic feet cargo space.

Lumber tally—The scale of the log after it is cut into lumber.

Mill run—All of the lumber without regard to grade which has a shipping value.

Moulding—The narrow strips of lumber moulded in various patterns used in interior finish.

Odd lengths—Lumber trimmed to odd feet, 7, 9, 11, 13, 15.

Off-bearer—The man who stands behind the saw and takes away the lumber and slabs. Syn.—Tail sawyer, swamper.

Ogee moulding—One having a double curve formed by a concave and convex line.

Overrun—The difference between the lumber tally and mill scale.

Parbuckle—A device for loading logs. It consists of a chain or rope in the form of a sling attached to the wagon or sled.

Partition—Lumber used for interior partitions where both sides of the board are exposed.

Pile bottom—The foundation (timber or concrete) on which lumber is piled.

Pin knot—A knot which is sound and not more than one-half inch in diameter.

Pitch—The angle between the back of a tooth and a line drawn from a point of the tooth to the back of a band saw or to the center of a circular saw.

Pitch pocket—An opening or space between the annual rings containing pitch in coniferous trees. Syn.—Pitch seams.

Plain sawed—All lumber which is not quarter sawed. Syn.—Flat grain, bastard grain, slash grain.

Pony gang—A saw crew of two men who do their own swamping.

Porch decking—A tongue and grooved board, the upper face of which is grooved in order to carry off rain water.

Quarter-sawed—In hardwoods, when the lumber is cut parallel or nearly so with the medullary rays. In soft woods, when the growth rings do not tip more than 45° from the vertical throughout the entire length of the board. Syn.—Center sawed, comb grained, edge grained, figure grained, rift sawed, silver grained, vertical grained.

Rip—To cut a board lengthwise.

Rock pine—The Chicago name for hemlock.

Sampson—An upright lever used for rolling logs. It is attached to the log by a chain.

Short length—Lumber from 4 to 10 feet.

Sidings—Boards sawed from the outer portion of a log when the central part is made into lumber.

Slack cooperage—Containers for nonliquid products consisting of two round heads and a body composed of staves held together with hoops.

Slat—A sawed piece of wood 7½ by 2½ by ¼ inches, used in pencil manufacture.

Small knot—A sound knot not more than three-fourths inch in diameter.

Snake—To make a wavy cut in sawing.

Snipping—The act of rounding off the end of a log to prevent the latter from sticking in the ground. Syn.—Sniping.

Sound knot—One which is as solid and hard across its face as the surrounding wood and firm enough to retain its place in the board.

Spike knot—A knot sawed lengthwise in the board.

Spline—A rectangular strip of wood which is substituted for the tongue on heavy factory flooring.

Stepping—A grade of lumber worked to a size suitable for steps.

Stock boards—Of even widths, usually 8, 10, and 12 inches.

Straight grain—A piece of lumber is said to be straight grained when the principal wood cells are parallel to its length.
Swamp hook—A hook to be attached to a chain used for rolling logs.
Tail sawyer—Off-bearer.
Tally—A record of the number of pieces and grades which are cut in the mill.
Tight cooperage—Containers for liquids consisting of two round heads and a body composed of several staves held together by hoops in such a manner as to hold liquids.
Timber—4 inches by 4 inches, and larger dimensions.
To box a log—To throw a log from the log trough upon the mill deck by means of a log kicker.
To box the heart—To cut boards from all sides of the heart, leaving the latter as a piece of timber.
To jack lumber—Means to pass up boards to the piler on top of the pile by leverage on an upright pole or a short board projecting from the front of the pile.
To saw around a log—To cut three or more faces on a log in order to get the best quality of lumber in each cut.
To saw alive—To make all cuts on the log parallel, without canting the log.
Torn grain—A defect on surfaced lumber caused by the fibers of the wood being torn by the planer knives.
Uppers—Finish lumber.
Veneer—A thin piece of lumber cut on a veneer machine. There are three kinds of veneer, viz, sawed, sliced, and rotary cut.
Wane—Bark or decrease in wood on the edge of board, plank, or timber.
Washboard lumber—Poorly sawed lumber with ridges on the face of the boards.
Woods scale—The scale of the logs made in the woods.
Wood fiber—Narrow shavings cut from a round block of wood by a special machine.
Yard lumber—Lumber which has been air dried.

ABBREVIATIONS.

C. I. F.—Cost, insurance, and freight.
F. O. B.—Free on board.
F. A. S.—Free along side.
F. G.—Flat grain.
V. G.—Vertical grain.
C. I. F. E.—Cost, insurance, freight, and exchange.
D. & H.—Dressed and headed. A flooring strip which has been surfaced, tongued, and grooved on one side, and also has a tongue on one end and a groove on the other, so that the joints may not necessarily come over a joint.
D. & M.—Dressed-and-matched boards, which have been tongued, grooved, and matched.
E. G.—Edge grain.
W. A. L.—Wider, all lengths.
AW. & AL.—All widths and all lengths.
5/4, 6/4, 8/4—1¼ inches, 1½ inches, and 2 inches.
K. D.—Kiln dried or knocked down.
S. 4S. C. S.—Surfaced on 4 sides in the 1/16-inch caulking seam on each edge.
S. M.—Surface measure.
S1 S. 1E—Surfaced on one side and one edge.
T. B. & S.—Top, bottom, and sides.
T. & G.—Tongued and grooved.

○

Spotted something that isn't quite right?

Let us know and we'll fix it for future reprints. And as a thank you for your diligence, we'll send you a free Sothis book of your choice.

publisher@sothis.press

This page intentionally left blank

Made in United States
Troutdale, OR
07/26/2025

33185208R00046